TEST BANK

FOR
PRESS AND SIEVER'S

UNDERSTANDING EARTH

SECOND EDITION

Simon M. Peacock
Sondra S. Peacock
Arizona State University

W. H. Freeman and Company
New York

ACQUISITIONS EDITOR: Patrick Shriner
ASSISTANT EDITOR: Melina LaMorticella
PROJECT EDITOR: Erica Lane Seifert
PRODUCTION COORDINATOR: Paul Rohloff
EDITORIAL ASSISTANTS: Maia Holden, Jennifer King
COMPOSITION AND TEXT DESIGN: Christopher Wieczerzak
MANUFACTURING: RR Donnelley & Sons Company

ISBN: 0-7167-2795-1

Printed in the United States of America

First Printing 1998

Table of Contents

Preface

From the Authors

As an aid in preparing introductory geology quizzes and exams, this *Test Bank* offers 1,200 multiple-choice questions (50 questions per chapter) for the second edition of Frank Press and Raymond Siever's *Understanding Earth*. We have attempted to present a variety of different types of questions including geologic definitions (vocabulary), historical questions, concept questions, and practical questions. Each chapter contains several sets of questions connected to geologic diagrams, such as cross sections. Several questions in the *Test Bank* can be modified to fit your local geology. All of the questions in the Test Bank are tied directly to specific pages in the textbook. We encourage instructors to modify *Test Bank* questions and answers in order to incorporate topics and examples presented in their lectures that are not discussed in the textbook. Instructors should note that the wording of some *Test Bank* questions may provide answers or clues to other questions.

The *CD-ROM* that accompanies the second edition of *Understanding Earth* contains practice multiple-choice questions (self-quizzes) with built-in feedback. These quizzes are designed to help students better understand the material and prepare for exams. Additional self-quizzes appear on the *Understanding Earth Web site*. Many of the *CD-ROM* and *Web site* self-quizzes are modified forms of questions presented in the first edition of the *Test Bank*. All of the *CD-ROM* and *Web site* self-quizzes contain different questions than those in the second edition of the *Test Bank*, although the questions cover similar concepts. Instructors may wish to add questions from the *CD-ROM* and *Web site* quizzes to their own exams in order to encourage students to use the self-quizzes.

In preparing this *Test Bank*, we have attempted to write clear, unambiguous questions based on our experience preparing such exams over the past decade. Despite our "experience," however, some of the questions that appear perfectly clear to us will strike students as unclear or ambiguous. If large numbers of students choose the same, incorrect answer to a question, then the instructor should scrutinize the question to see if the wording was confusing, and if so, modify the answer key or eliminate the question from the exam.

At many universities, introductory geology lectures involve presenting lectures to hundreds of students at a time. Because of large class sizes, exams commonly consist of multiple-choice questions graded by computer scanning. Nevertheless, in addition to multiple-choice questions, we strongly recommend that all introductory geology exams include one or two short essays to test and strengthen the student's ability to present ideas clearly in writing. The essay grading burden can be significantly reduced by (1) grading essays on a coarse 0-2-4-6-8-10 point scale, so the instructor does not waste time worrying about differences involving one point, and (2) presenting examples of good essays in class. The latter technique frees the instructor from making detailed corrections on each essay and also demonstrates the range of student responses that merit full credit.

Simon M. Peacock
Sondra S. Peacock
July, 1997

1

Introduction

1.   Living organisms have been on Earth for _____ of Earth's history?
        A.   less than 1%
        B.   about 20%
        C.   about 50%
        D.   about 80%
Answer = D (page 4)

2.   Humans have been on Earth for _____ of Earth's history.
        A.   less than 1%
        B.   approximately 5%
        C.   approximately 50%
        D.   approximately 90%
Answer = A (page 4)

3.   Which of the following statements regarding the scientific method is true?
        A.   A hypothesis must be agreed upon by more than one scientist.
        B.   A theory is a hypothesis that has withstood many scientific tests.
        C.   A theory is proven to be true, and therefore may not be discarded.
        D.   A hypothesis cannot predict the outcome of scientific experiments.
Answer = B (page 4)

4.   According to the principle of uniformitarianism, _____ .
        A.   geologic processes we observe today have operated in the past
        B.   humans evolved from apes
        C.   all of the planets formed from a uniform solar nebula
        D.   early Earth was covered by a uniform magma ocean
Answer = A (page 4)

5.   Approximately how long ago did the Big Bang take place?
    A.   10–15 thousand years ago
    B.   10–15 million years ago
    C.   100–150 million years ago
    D.   10–15 billion years ago
Answer = D (page 4)

6.   What are the two most abundant elements in nebulae (gas clouds) in the universe?
    A.   oxygen and nitrogen
    B.   oxygen and silicon
    C.   hydrogen and helium
    D.   carbon and silicon
Answer = C (page 5)

7.   Under intense pressure and high temperature, hydrogen atoms combine to form helium. This process is called _____.
    A.   nuclear fusion
    B.   metamorphism
    C.   evolution
    D.   convection
Answer = A (page 6)

8.   What caused dust and condensing material to accrete into planetesimals?
    A.   heating of gases
    B.   gravitational attraction and collisions
    C.   nuclear fusion
    D.   rotation of the proto–sun
Answer = B (page 6)

9.   Which of the following is not one of the four inner planets?
    A.   Mars
    B.   Earth
    C.   Mercury
    D.   Neptune
Answer = D (page 6)

10.   The giant outer planets are composed mostly of _____?
    A.   rocks and ice
    B.   oxygen and nitrogen
    C.   hydrogen and helium
    D.   helium and ice
Answer = C (page 7)

11. The process by which an originally homogeneous Earth developed a dense core and a light crust is called _____ .
    A. metamorphism
    B. differentiation
    C. accretion
    D. compression
Answer = B (page 8)

12. The Moon is _____.
    A. older than the sun
    B. older than meteorites
    C. older than the Earth
    D. none of these
Answer = D (page 9)

13. The heat that caused melting in Earth's early history was supplied from which of the following events or causes?
    A. volcanic activity and radioactivity
    B. solar heating and volcanic activity
    C. a large impact event and radioactivity
    D. a large impact event and solar heating
Answer = C (page 9)

14. How thick is the Earth's crust?
    A. about 4 m
    B. about 4 km
    C. about 40 km
    D. about 400 km
Answer = C (page 10)

15. The name of the layer in the Earth that separates the crust from the core is called the _____ .
    A. magma
    B. lithosphere
    C. mantle
    D. continent
Answer = C (page 10)

16.   The inner core of the earth is solid because _____ .
        A.   pressures are high in the inner core
        B.   temperatures are low in the inner core
        C.   the composition of the inner core is similar to that of the Earth's solid crust
        D.   None of these; the inner core is liquid.
Answer = A (page 10)

17.   Ninety percent of the Earth is made up of which four elements?
        A.   iron, oxygen, silicon, and magnesium
        B.   oxygen, nitrogen, hydrogen, and silicon
        C.   magnesium, aluminum, silicon, and oxygen
        D.   silicon, calcium, aluminum, and iron
Answer = A (page 10)

18.   Fifty percent of the Earth's crust is made up of which element?
        A.   iron
        B.   silicon
        C.   oxygen
        D.   aluminum
Answer = C (page 10)

19.   What are the two most abundant elements (by weight) in the whole Earth?
        A.   oxygen and silicon
        B.   silicon and sodium
        C.   oxygen and iron
        D.   iron and silicon
Answer = C (pages 10–11)

20.   Oxygen makes up a large proportion of the Earth's _____ .
        A.   atmosphere
        B.   oceans
        C.   crust
        D.   atmosphere, oceans, and crust
Answer = D (pages 10–11)

21.   What powers the Earth's internal heat engine?
        A.   radioactivity
        B.   solar energy
        C.   volcanoes
        D.   ocean tides
Answer = A (page 11)

22. The Earth's external heat engine is not responsible for which of the following:
    A. climate
    B. erosion
    C. tides
    D. wind
Answer = C (page 11)

23. Which of the following gases is not expelled from volcanoes today?
    A. carbon dioxide
    B. water vapor
    C. nitrogen
    D. oxygen
Answer = D (page 11)

24. Which gases released from volcanoes 4 billion years ago are not released today?
    A. nitrogen
    B. water vapor
    C. hydrogen
    D. The gases released 4 billion years ago are the same as those released today.
Answer = D (page 11)

25. Oxygen built up in the Earth's atmosphere because _____.
    A. the oceans separated from the crust
    B. rocks weathered and released their oxygen
    C. algae and other organisms employed photosynthesis
    D. oxygen settled on Earth from planets further from the sun
Answer = C (page 11)

26. Which of the following planets is not geologically active?
    A. Mars
    B. Mercury
    C. Venus
    D. Earth
Answer = B (page 12)

27. The evidence that Mars once had water includes _____.
    A. dry river beds and valleys
    B. large salt deposits from dried-up oceans
    C. liquid water in current Martian oceans
    D. water vapor in the Martian atmosphere
Answer = A (page 13)

28.   The differentiation of the Moon resulted in _____.
       A.   a core, mantle, and crust similar to the Earth
       B.   a large mantle and a thin crust
       C.   a small core and a thick crust.
       D.   None of the above; the Moon did not undergo differentiation
       Answer = C (page 13)

29.   When did geologists develop the theory of plate tectonics?
       A.   in the mid-1800s
       B.   in the 1890s
       C.   in the 1940s
       D.   in the 1960s
       Answer = D (page 14)

30.   The lithosphere is approximately ____ km thick.
       A.   1–2
       B.   5–10
       C.   50–100
       D.   500–1000
       Answer = C (page 14)

31.   The asthenosphere is _____.
       A.   cool and strong
       B.   cool and weak
       C.   hot and strong
       D.   hot and weak
       Answer = D (page 14)

32.   The Earth's lithosphere is broken into approximately _____ large, rigid plates.
       A.   2
       B.   12
       C.   50
       D.   200
       Answer = B (page 15)

33.   Which of the following statements about convection is true?
       A.   Heat is transferred from hot material to cool material without inducing
             a flow.
       B.   Hot material flows upward and displaces cool material.
       C.   Cool material flows upward and displaces hot material.
       D.   Random circulation occurs.
       Answer = B (page 15)

34. Approximately how fast does the Earth's lithospheric plate move?
    A. several centimeters per year
    B. several centimeters per day
    C. several centimeters per hour
    D. several centimeters per second
    Answer = A (page 16)

35. Which of the following is not a type of plate boundary?
    A. convergent plate boundary
    B. transform fault plate boundary
    C. divergent plate boundary
    D. all of these are plate boundaries
    Answer = D (page 17)

36. The following picture illustrates which type of plate boundary?

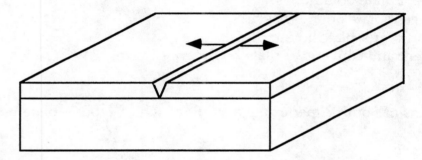

    A. transform
    B. strike-slip
    C. convergent
    D. divergent
    Answer = D (page 16)

37. New seafloor is created at a _____.
    A. deep sea trench
    B. mid-ocean ridge
    C. subduction zone
    D. transform fault
    Answer = B (page 17)

38. The descent of oceanic lithosphere into the mantle is the process of _____ .
    A. accretion
    B. subduction
    C. divergence
    D. contraction
    Answer = B (page 17)

39. Approximately how deep (below sea level) are deep-sea trenches?
    A. 1 km
    B. 10 km
    C. 100 km
    D. 1000 km
Answer = B (page 17)

40. Which of the following features is not associated with a convergent plate boundary?
    A. a mid-ocean ridge
    B. earthquakes
    C. a deep-sea trench
    D. volcanic activity
Answer = A (page 17)

41. Volcanism is associated with which of the following types of plate boundaries?
    A. divergent plate boundaries
    B. convergent plate boundaries
    C. transform fault plate boundaries
    D. divergent and convergent plate boundaries
Answer = D (page 17)

42. The Andes Mountains of South America are a result of which type of plate boundary?
    A. convergent
    B. divergent
    C. transform
    D. They are not related to a plate boundary.
Answer = A (page 17)

43. Mount St. Helens, a part of the Cascade Range of volcanoes, results from the subduction of which plate?
    A. Pacific Plate
    B. Cocos Plate
    C. Juan de Fuca Plate
    D. North American Plate
Answer = C (page 17)

44. What is the name of the large supercontinent that existed 200 million years ago when all of the continents were together?
    A. San Andreas
    B. Andes
    C. Pangaea
    D. Nebulae
Answer = C (page 19)

45. A geologist could be involved in which of the following projects.
    A. designing flood control systems
    B. mapping the ocean floor
    C. finding natural gas reserves
    D. all of the above
    Answer = D (page 20)

46. Why is our vulnerability to natural disasters growing?
    A. because the frequency of volcanic eruptions is increasing
    B. because the human population is increasing
    C. because the number of earthquakes each year is increasing
    D. because the number of floods each year is increasing
    Answer = B (page 21)

47. Which of the following is not an aspect of geology?
    A. global environment
    B. biotechnology
    C. natural disasters
    D. economic geology
    Answer = B (page 21)

48. The greatest natural threat to our environment is _____.
    A. humans
    B. volcanoes
    C. oceans
    D. bacteria
    Answer = A (page 21)

49. Which of the following was not considered evidence that continents move around the Earth's surface?
    A. The outlines of South America and Africa match like a jigsaw puzzle.
    B. Fossils and living organisms are similar between Australia and India.
    C. Rock formations from North America and Europe match.
    D. Volcanoes rim the Pacific Ocean.
    Answer = D (page 22)

50. The theory of plate tectonics was initially not widely accepted because _____.
    A. land bridges would have blocked plate movement
    B. rocks of the outer Earth were considered too stiff for continents to move through them
    C. fossils on South America and Africa did not match
    D. ocean floor mapping showed that older rocks occur away from mid-ocean ridges
    Answer = B (page 23)

2

Minerals:
Building Blocks of Rocks

1. Solid materials that do not possess an orderly arrangement of atoms are called _____ .
   A. glasses
   B. minerals
   C. crystals
   D. polymorphs
   Answer = A (page 28)

2. Each element has a unique number of _____ .
   A. protons
   B. neutrons
   C. electrons
   D. all of these
   Answer = A (page 30)

3. The atomic number for silicon is 14, meaning that oxygen atoms have _____ .
   A. 14 protons
   B. 14 neutrons
   C. 7 protons and 7 neutrons
   D. none of these
   Answer = A (page 30)

4.   The atomic weight of an element is equal to _____.
      A.   the number of protons
      B.   the number of neutrons
      C.   the number of protons and neutrons
      D.   none of these
Answer = C (page 30)

5.   Atoms that have gained or lost electrons are no longer electrically neutral and are called _____ .
      A.   ions
      B.   polymorphs
      C.   electrons
      D.   isotopes
Answer = A (page 31)

6.   What does the symbol Ca^+ stand for?
      A.   a calcium electron
      B.   a calcium proton
      C.   a calcium cation
      D.   a calcium anion
Answer = C (page 31)

7.   In the periodic table of the elements, as one goes from left to right along a row _____ .
      A.   the number of electrons in the outer shell increases
      B.   the atomic number decreases
      C.   the number of isotopes increases
      D.   all of these
Answer = A (page 32)

8.   Which of the following elements has a strong tendency to lose electrons from the outermost electron shell?
      A.   chlorine
      B.   oxygen
      C.   sodium
      D.   sulfur
Answer = C (page 33)

9.   Which of the following elements commonly forms negative ions?
      A.   hydrogen
      B.   magnesium
      C.   oxygen
      D.   sodium
Answer = C (page 34)

10. What is the dominant type of bonding in minerals?
    A. covalent bonding
    B. ionic bonding
    C. metallic bonding
    D. nuclear bonding
    Answer = B (page 35)

11. Diamond is an example of what type of bonding?
    A. covalent
    B. ionic
    C. metallic
    D. nuclear
    Answer = A (page 35)

12. The growth of a solid from a material whose atoms can come together in the proper chemical proportions and crystalline arrangement is called _____ .
    A. density
    B. bonding
    C. melting
    D. crystallization
    Answer = D (page 36)

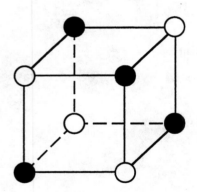

13. What elements could be represented by the open and solid spheres in the mineral structure depicted above?
    A. carbon and oxygen
    B. iron and magnesium
    C. silicon and oxygen
    D. sodium and chlorine
    Answer = D (page 36)

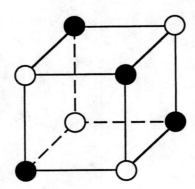

14. Which of the following minerals could be represented by the mineral structure depicted above?
     A.   galena
     B.   graphite
     C.   olivine
     D.   quartz
Answer = A (page 36)

15. Large crystals with well-formed crystal faces tend to form when ____ .
     A.   molten rock cools quickly
     B.   rocks undergo melting
     C.   minerals have the space to grow such as in open cavities
     D.   volcanoes erupt explosively
Answer = C (page 37)

16. Which of the following elements tends to form positively-charged ions rather than negatively-charged ions?
     A.   chlorine
     B.   iron
     C.   oxygen
     D.   sulfur
Answer = B (page 38)

17. Iron and magnesium ions are similar in size and both have a "+2" positive charge. Therefore, we would expect _____ .
     A.   iron and magnesium to bond easily
     B.   iron and magnesium to share electrons
     C.   iron and magnesium to be polymorphs
     D.   iron and magnesium to substitute for each other in minerals
Answer = D (page 39)

18.   Chemical substances that have exactly the same chemical formula, but different crystal structures, are called _____ .
      A.   ions
      B.   polymorphs
      C.   electrons
      D.   isotopes
Answer = B (page 39)

19.   Which of the following statements about graphite and diamond is false?
      A.   Graphite and diamond have the same density.
      B.   Graphite and diamond have different mineral structures.
      C.   Graphite and diamond are both made of carbon atoms.
      D.   Graphite is stable in the Earth's crust whereas diamond is stable in the Earth's mantle.
Answer = A (page 39)

20.   The chemical formula $(Mg,Fe)_2SiO_4$ describes which of the following minerals?
      A.   feldspar
      B.   mica
      C.   olivine
      D.   pyroxene
Answer = C (page 39)

21.   Most common rock-forming minerals are _____ .
      A.   carbonates
      B.   oxides
      C.   silicates
      D.   sulfides
Answer = C (page 40)

22.   The two most common elements in the Earth's crust are:
      A.   calcium and carbon
      B.   chlorine and sodium
      C.   iron and sulfur
      D.   silicon and oxygen
Answer = D (page 40)

23.   Which of the following structures best depicts a silicate ion?

(A)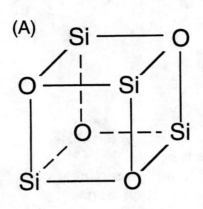

(B)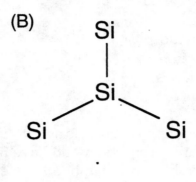

(C)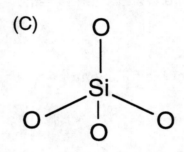

(D)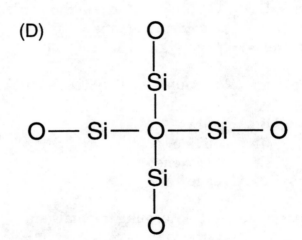

Answer = C (page 41)

24.   The mineral pyroxene is an example of a _____ silicate.
       A.   framework
       B.   single chain
       C.   sheet
       D.   ring
Answer = B (page 41)

25.   Which mineral consists of isolated silica tetrahedra?
       A.   calcite
       B.   olivine
       C.   pyroxene
       D.   quartz
Answer = B (page 41)

26. What type of mineral is calcite?
    A. carbonate
    B. oxide
    C. silicate
    D. sulfide
    Answer = A (page 42)

27. Clay minerals are common examples of _____ silicates.
    A. framework
    B. single chain
    C. sheet
    D. ring
    Answer = C (page 42)

28. Which of the following minerals is a common clay mineral used for making pottery?
    A. feldspar
    B. kaolinite
    C. olivine
    D. pyroxene
    Answer = B (page 42)

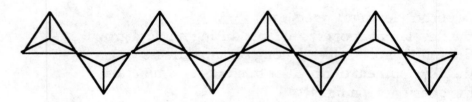

29. What silicate mineral contains tetrahedra linked together as depicted in the diagram above?
    A. mica
    B. olivine
    C. pyroxene
    D. quartz
    Answer = C (pages 41–42)

30. Which of the following elements is never found in nature as a pure, native element?
    A. carbon
    B. copper
    C. gold
    D. silicon
    Answer = D (page 43)

31.   What type of mineral is pyrite, also known as "fool's gold"?
       A.   silicate
       B.   carbonate
       C.   oxide
       D.   sulfide
    Answer = D (page 44)

32.   What element occurs in quartz, olivine, and calcite?
       A.   oxygen
       B.   silicon
       C.   magnesium
       D.   calcium
    Answer = A (page 42)

33.   Which of the following minerals has the greatest hardness?
       A.   corundum
       B.   feldspar
       C.   quartz
       D.   talc
    Answer = A (page 46)

34.   The mineral property "cleavage" refers to _____.
       A.   the development of crystal faces during mineral growth
       B.   the splitting of a mineral along planar surfaces
       C.   the development of irregular fractures in a mineral
       D.   the density of a mineral
    Answer = B (page 47)

35.   Which of the following minerals has three excellent cleavage planes at right angles to each other?
       A.   calcite
       B.   halite
       C.   mica
       D.   quartz
    Answer = B (page 48)

36.   Which of the following minerals does not exhibit cleavage?
       A.   quartz
       B.   amphibole
       C.   calcite
       D.   mica
    Answer = A (page 49)

37.   Which of the following mineral properties is the least reliable clue to its identity?
      A.   hardness
      B.   cleavage
      C.   color
      D.   density
Answer = C (page 50)

38.   Which of the following statements about asbestos is false?
      A.   Only one asbestos mineral, crocidolite, forms sharp fibers.
      B.   Most asbestos minerals are amphiboles.
      C.   U.S. government regulations only apply to the asbestos mineral crocidolite.
      D.   Several fatal lung diseases have been linked to inhaling certain kinds of asbestos.
Answer = C (page 52)

39.   The shape in which an individual crystal grows is called the mineral's crystal _____.
      A.   cleavage
      B.   habit
      C.   density
      D.   streak
Answer = B (page 53)

40.   Which of the following is not a mineral property?
      A.   isotope
      B.   hardness
      C.   cleavage
      D.   density
Answer = A (page 51)

41.   Which of the following statements regarding the density of minerals is false?
      A.   Density depends on the atomic weight of the ions in a mineral.
      B.   Density depends on the closeness of the atomic packing.
      C.   Density increases with increasing pressure.
      D.   Density increases with increasing temperature.
Answer = D (page 51)

42.   Most minerals are classified on the basis of _____ .
      A.   the arrangement of silica tetrahedra
      B.   their anions
      C.   their color
      D.   their density
Answer = B (page 40)

43.   Given two minerals with the exact same chemical composition, which mineral is more likely to form at higher pressure?
      A.   the mineral with the higher hardness
      B.   the mineral with the lower hardness
      C.   the mineral with the higher density
      D.   the mineral with the lower density
Answer = C (page 39)

44.   Given two minerals with the exact same chemical composition, which mineral is more likely to form at higher temperature?
      A.   the mineral with the higher hardness
      B.   the mineral with the lower hardness
      C.   the mineral with the higher density
      D.   the mineral with the lower density
Answer = D (page 39)

45.   Carbon has an atomic number of 6 and an atomic weight of 12.011.This means that _____.
      A.   carbon atoms have 6 protons and 12.011 neutrons
      B.   carbon atoms have 6 protons and a density of 12.011 grams per cubic centimeter
      C.   carbon atoms have 6 neutrons and 12.011 protons
      D.   carbon atoms have 6 protons and varying numbers of neutrons
Answer = D (page 30)

46.   Which of the following substances is considered a mineral?
      A.   seawater
      B.   rock salt
      C.   cast iron
      D.   vegetation
Answer = B (page 28)

47.   The chemical symbol "Fe" refers to _____ .
      A.   the mineral pyrite
      B.   the mineral feldspar
      C.   the element sodium
      D.   the element iron
Answer = D (page 33)

48.   Which of the following minerals will have a hexagonal cross-section when cut at right angles to the long dimension of the crystal?
      A.   halite
      B.   galena
      C.   quartz
      D.   diamond
Answer = C (page 37)

49.   Which of the following metals is commonly mined as an oxide?
      A.   iron
      B.   nickel
      C.   copper
      D.   zinc
Answer = A (page 44)

50.   Which of the following statements is false?
      A.   Pyroxenes and amphiboles are both silicates.
      B.   Pyroxenes and amphiboles have cleavages at different angles.
      C.   Pyroxenes and amphiboles have the same chemical composition.
      D.   Pyroxenes and amphiboles consist of chains of silica tetrahedra.
Answer = C (page 41)

3

Rocks:

Records of Geologic Processes

1.   What two properties define a rock?
    A.   texture and mineralogy
    B.   subsidence and uplift
    C.   outcrops and streams
    D.   crystallization and precipitation
Answer = A (page 60)

2.   Mineral grains in most rocks are several _____ in diameter.
    A.   millimeters
    B.   centimeters
    C.   meters
    D.   kilometers
Answer = A (page 60)

3.   Which of the following is not one of the three major rock types?
    A.   igneous
    B.   mineralogic
    C.   sedimentary
    D.   metamorphic
Answer = B (page 60)

4.   Igneous rocks form by _____.
     A.   crystallization of molten rock
     B.   solid state changes due to increased temperatures and/or pressures
     C.   lithification of sediments
     D.   chemical or biochemical precipitation of minerals
     Answer = A (page 60)

5.   Metamorphic rocks form by _____.
     A.   crystallization of molten rock
     B.   solid state changes due to increased temperatures and/or pressures
     C.   lithification of sediments
     D.   chemical or biochemical precipitation of minerals
     Answer = B (page 60)

6.   Clastic sedimentary rocks form by _____.
     A.   crystallization of molten rock
     B.   solid state changes due to increased temperatures and/or pressures
     C.   lithification of sediments
     D.   chemical or biochemical precipitation of minerals
     Answer = C (page 60)

7.   Oil forms _____ .
     A.   in volcanic rocks that have abundant feldspar
     B.   in contact metamorphic rocks
     C.   in sedimentary rocks that are rich in organic remains
     D.   in plutonic rocks that are rich in CaO
     Answer = C (page 61)

8.   Volcanic rocks have smaller crystals than plutonic rocks because _____ .
     A.   plutonic rocks cool slower than volcanic rocks
     B.   volcanic rocks cool slower than plutonic rocks
     C.   plutonic rocks melt slower than volcanic rocks
     D.   volcanic rocks melt slower than plutonic rocks
     Answer = A (pages 61–62)

9.   Clastic sedimentary rocks are lithified by _____ .
     A.   compaction
     B.   cementation
     C.   both compaction and cementation
     D.   neither compaction nor cementation
     Answer = B (page 62)

10. Which of the following properties is diagnostic of a sedimentary rock?
    A. bedding
    B. foliation
    C. glassy texture
    D. orogeny
    Answer = A (page 62)

11. Which of the following processes is not involved in hardening sediments into sedimentary rocks?
    A. cementation
    B. compaction
    C. metamorphism
    D. All of these are involved
    Answer = C (page 62)

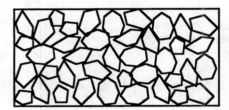

Grains are plagioclase and quartz
Area surrounding grains is open space

1 cm   scale

12. The diagram above most likely represents which type of rock?
    A. extrusive igneous
    B. metamorphic
    C. clastic sedimentary
    D. biochemical sedimentary
    Answer = C (page 62)

13. Which of the following minerals is not commonly found in sedimentary rocks?
    A. amphibole
    B. feldspar
    C. dolomite
    D. gypsum
    Answer = A (page 62)

14. Which of the following minerals is not commonly found in metamorphic rocks?
    A. quartz
    B. garnet
    C. pyroxene
    D. olivine
    Answer = D (pages 62, 64)

15.   Chemical sediments form from _____.
        A.   rapid cooling of molten sediments and calcium carbonate shells
        B.   particles of sediment deposited in water
        C.   precipitation of minerals from sea water
        D.   accumulation of calcium carbonate shells
    Answer = C (page 62)

16.   Which of the following minerals is not commonly found in igneous rocks?
        A.   quartz
        B.   feldspar
        C.   mica
        D.   staurolite
    Answer = D (page 62)

17.   Halite, a rock that forms from precipitation of sodium chloride, is an example of a(n) _____.
        A.   extrusive igneous rock
        B.   intrusive igneous rock
        C.   clastic sedimentary rock
        D.   chemical sedimentary rock
    Answer = D (page 62)

18.   Which of the following does not form from molten rock?
        A.   volcanic ash
        B.   limestone
        C.   basalt
        D.   granite
    Answer = B (page 62)

19.   Which of the following minerals commonly occurs in igneous, metamorphic, and sedimentary rocks?
        A.   calcite
        B.   olivine
        C.   feldspar
        D.   mica
    Answer = C (page 62)

20.   Which of the following rocks is best described as a foliated metamorphic rock?
        A.   sandstone
        B.   granite
        C.   basalt
        D.   schist
    Answer = D (page 62)

21.   Which of the following processes can deposit clastic sediments?
      A.   ice
      B.   wind
      C.   running water
      D.   all of these
Answer = D (page 62)

22.   Which of the following statements is false?
      A.   Extrusive igneous rocks generally contain fine-grained crystals and/or glass.
      B.   Extrusive and intrusive igneous rocks may contain feldspar, pyroxene, and mica.
      C.   Plutonic rocks cool slowly from a magma.
      D.   Igneous rocks cover most of the surface of the continents.
Answer = D (pages 62–63)

23.   What are the most abundant group of minerals in clastic sedimentary rocks?
      A.   oxides
      B.   carbonates
      C.   silicates
      D.   sulfates
Answer = C (page 63)

24.   Which of the following rock types is the least abundant in the Earth's crust?
      A.   igneous
      B.   sedimentary
      C.   metamorphic
      D.   Igneous, sedimentary, and metamorphic rocks are equally abundant.
Answer = B (page 63)

25.   Which of the following textures is most common in a sedimentary rock?
      A.   granular
      B.   foliated
      C.   bedded
      D.   striated
Answer = C (page 63)

26.   Metamorphic rocks are produced by mineralogical and textural changes of which type of rock?
      A.   igneous
      B.   sedimentary
      C.   metamorphic
      D.   all of these
Answer = D (page 64)

27.   Which of the following statements regarding metemorphic rocks is true?
    A.   Metamorphic rocks undergo partial melting to change their mineralogy.
    B.   Metamorphic rocks must form very deep in the Earth's mantle where pressures and temperatures are high.
    C.   All metamorphic rocks exhibit foliation, which is a characteristic and distinguishing feature.
    D.   Metamorphic rocks are commonly found at convergent plate boundaries and near igneous intrusions.
Answer = D (page 64)

28.   Which of the following properties is diagnostic of a regionally metamorphosed rock?
    A.   bedding
    B.   foliation
    C.   glassy texture
    D.   orogeny
Answer = B (page 64)

29.   Regional metamorphism occurs _____.
    A.   next to igneous intrusions
    B.   during plate collisions
    C.   at the Earth's surface
    D.   during deposition of sediments
Answer = B (page 64)

30.   A typical basalt consists of approximately two-thirds _____.
    A.   calcium, magnesium, and sodium oxides
    B.   silicon and aluminum oxides
    C.   iron and potassium oxides
    D.   silicon oxide
Answer = B (page 65)

31.   Which of the following is not a major oxide in igneous rocks?
    A. Al_2O_3
    B. CO_2
    C. CaO
    D. FeO
Answer = B (page 65)

32. Which of the following statements about water (H_2O) is false?
    A. Clay minerals hold water in sedimentary rocks.
    B. Some rocks contain up to 5% water.
    C. Igneous rocks contain more water than sedimentary rocks.
    D. Sedimentary rocks contain water in the pore spaces.
    Answer = C (page 65)

33. To date, the deepest drill hole reaches a depth of approximately _____.
    A. 1 km
    B. 12 km
    C. 240 km
    D. 4000 km
    Answer = B (page 66)

34. Places where bedrock is exposed at the Earth's surface are called _____.
    A. outcrops
    B. sediments
    C. subsidences
    D. intrusives
    Answer = A (page 66)

35. Outcrops are scarce in which of the following locations in the U.S.?
    A. Pacific Coast
    B. Rocky Mountains
    C. Midwest
    D. Appalachian Mountains
    Answer = C (page 66)

36. In a typical drill hole located in the American Midwest, we would expect to find _____.
    A. several kilometers of metamorphic rocks overlying igneous rocks
    B. several kilometers of igneous rocks overlying metamorphic rocks
    C. several kilometers of igneous and metamorphic rocks overlying sedimentary rocks
    D. several kilometers of sedimentary rocks overlying igneous and metamorphic rocks
    Answer = D (page 66)

37. Which of the following types of rocks form in the interior of the Earth?
    A. clastic sedimentary rocks
    B. biochemical sedimentary rocks
    C. volcanic igneous rocks
    D. plutonic igneous rocks
    Answer = D (page 68)

38.   Slow crystallization of a magma is responsible for the texture of _____ rocks.
    A.   plutonic
    B.   volcanic
    C.   sedimentary
    D.   metamorphic
Answer = A (page 68)

39.   Which of the following statements about James Hutton is false?
    A.   He wrote *Theory of the Earth with Proof and Illustrations*.
    B.   He is considered the founder of plate tectonics.
    C.   He developed the concept of the rock cycle.
    D.   He recognized the cyclic nature of geologic change.
Answer = B (page 68)

40.   Which of the following mineral assemblages would not be found in an igneous rock?
    A.   quartz, feldspar, mica
    B.   pyroxene, feldspar, quartz
    C.   amphibole, pyroxene, feldspar
    D.   mica, garnet, pyroxene
Answer = D (page 69)

41.   Where does the rock cycle start?
    A.   crystallization of molten rock
    B.   deposition of sediment
    C.   metamorphism
    D.   There is no "start"—all rocks form from preexisting rocks.
Answer = D (page 69)

42.   At which of the following locations are igneous rocks least likely to form?
    A.   convergent plate boundaries
    B.   divergent plate boundaries
    C.   transform plate boundaries
    D.   hot spots
Answer = C (page 70)

43.   Which of the following statements about an orogeny is false?
    A.   An orogeny occurs during the collision of two plates.
    B.   An orogeny is unrelated to plate tectonics.
    C.   An orogeny involves mountain building.
    D.   An orogeny involves deformation processes.
Answer = B (page 69)

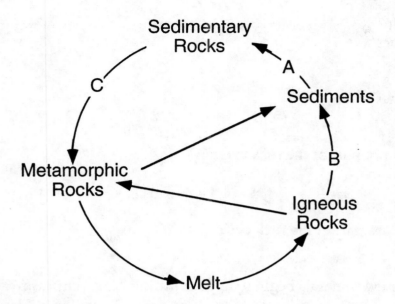

44. Referring to the diagram above, path A is _____.
    A.   cooling and crystallization
    B.   burial and lithification
    C.   weathering and deposition
    D.   cooling and uplift
    Answer = B (page 69)

45. Referring to the diagram above, what must occur along path B?
    A.   uplift, weathering and erosion, deposition
    B.   deposition, heat and pressure, weathering
    C.   melting, crystallization, heat and pressure
    D.   deposition, lithification, crystallization
    Answer = A (page 69)

46. Referring to the diagram above, what factor(s) are responsible for path C?
    A.   melting
    B.   crystallization
    C.   heat and pressure
    D.   burial and lithification
    Answer = C (page 69)

47. Subsidence of the Earth's crust allows _____.
    A.   lithification of sediments into rock
    B.   weathering of rock near the Earth's surface
    C.   melting of sediments
    D.   accumulation of additional layers of sediment
    Answer = D (page 69)

48.   What drives the rock cycle?
      A.   plate tectonics
      B.   gravity
      C.   tidal forces
      D.   electromagnetism
Answer = A (page 70)

49.   Which of the following is not part of the rock cycle?
      A.   metamorphism
      B.   melting
      C.   lithification
      D.   All of these are part of the rock cycle.
Answer = D (page 70)

50.   At what temperature do new minerals begin to appear during metamorphism?
      A.   30°C
      B.   100°C
      C.   300°C
      D.   1000°C
Answer = C (page 70)

4

Igneous Rocks:
Solids from Melts

1.   Which of the following best describes a granite?
      A.   A light-colored, fine-grained igneous rock rich in silica.
      B.   A light-colored, fine-grained igneous rock poor in silica.
      C.   A light-colored, coarse-grained igneous rock rich in silica.
      D.   A light-colored, coarse-grained igneous rock poor in silica.
Answer = C (page 81)

2.   Which of the following best describes a basalt?
      A.   A dark-colored, fine-grained igneous rock rich in silica.
      B.   A dark-colored, fine-grained igneous rock poor in silica.
      C.   A dark-colored, coarse-grained igneous rock rich in silica.
      D.   A dark-colored, coarse-grained igneous rock poor in silica.
Answer = B (page 82)

3.   Where would you expect to find the largest crystals in a lava flow?
      A.   near the top surface of the flow
      B.   in the center of the flow
      C.   near the bottom surface of the flow
      D.   the crystals would have the same grain size throughout
Answer = B (page 76)

4.   What geologist is credited with figuring out the origin of granite?
         A.   James Hutton
         B.   Friedrich Mohs
         C.   Charles Darwin
         D.   N. L. Bowen
     Answer = A (page 77)

5.   The famous Mother Lode of the 1849 gold rush in California is an example of
     _____.
         A.   a basaltic dike
         B.   volcanic ash
         C.   a hydrothermal vein
         D.   a granitic pluton
     Answer = C (page 98)

6.   Most igneous rocks contain _____ SiO_2 by weight.
         A.   less than 20%
         B.   between 40 and 70%
         C.   between 80 and 90%
         D.   100%
     Answer = B (page 79)

7.   Which of the following minerals is the most abundant mineral in ultramafic rocks?
         A.   amphibole
         B.   olivine
         C.   plagioclase feldspar
         D.   quartz
     Answer = B (page 82)

8.   Which of the following minerals is not commonly found in a mafic igneous rock?
         A.   quartz
         B.   plagioclase feldspar
         C.   potassium feldspar
         D.   olivine
     Answer = A (page 79)

9.   Which of the following igneous rocks has the same chemical composition as a granite?
         A.   gabbro
         B.   rhyolite
         C.   diorite
         D.   andesite
     Answer = B (page 81)

10. Which of the following igneous rocks has the same chemical composition as a basalt?
    A. gabbro
    B. rhyolite
    C. diorite
    D. andesite
    Answer = A (page 82)

11. Which of the following igneous rocks is the most mafic?
    A. diorite
    B. gabbro
    C. granite
    D. peridotite
    Answer = D (page 81)

12. In what type of igneous rock would you be most likely to find the mineral - $KAlSi_3O_8$?
    A. diorite
    B. gabbro
    C. granite
    D. peridotite
    Answer = C (page 79)

13. Which of the following statements about mafic rocks is true?
    A. Mafic rocks are richer in silica than felsic rocks.
    B. Mafic rocks crystallize at higher temperatures than felsic rocks.
    C. Mafic rocks are more viscous than felsic rocks.
    D. Mafic rocks tend to be lighter in color than felsic rocks.
    Answer = B (pages 81–82)

14. Which of the following igneous rocks crystallizes near the Earth's surface?
    A. basalt
    B. gabbro
    C. granite
    D. peridotite
    Answer = A (page 78)

15. What is the approximate silica content of a granite?
    A. 20 percent
    B. 45 percent
    C. 70 percent
    D. 100 percent
    Answer = C (page 81)

16.   Which mineral is commonly found in both felsic and mafic igneous rocks?
      A.   quartz
      B.   plagioclase feldspar
      C.   pyroxene
      D.   olivine
Answer = B (page 81)

17.   Which of the following rock types is not an extrusive igneous rock?
      A.   andesite
      B.   basalt
      C.   granite
      D.   rhyolite
Answer = C (page 78)

18.   What type of volcanic rock contains a large number of cavities (bubbles) that form when gases escape from the molten rock?
      A.   granite
      B.   obsidian
      C.   porphyry
      D.   pumice
Answer = D (page 78)

19.   What type of volcanic rock was used as arrowheads by Native Americans?
      A.   granite
      B.   obsidian
      C.   porphyry
      D.   pumice
Answer = B (page 78)

Felsic ⟶ Intermediate ⟶ Mafic

20.   Which of the following properties increases in the direction of the arrows in the diagram above?
      A.   melting temperature
      B.   potassium content
      C.   silica content
      D.   viscosity
Answer = A (page 82)

21.   What type of magma forms at mid-ocean ridges?
        A.   andesite
        B.   basalt
        C.   granite
        D.   peridotite
Answer = B (page 86)

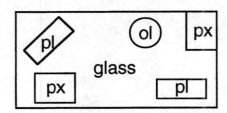

ol = olivine crystals
pl = plagioclase crystals
px = pyroxene crystals

scale
1 mm

22.   Which of the following rock types is depicted in the diagram above?
        A.   andesite
        B.   basalt
        C.   granite
        D.   peridotite
Answer = B (page 81)

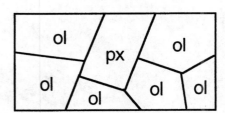

ol = olivine crystals
px = pyroxene crystals

scale
1 cm

23.   What type of igneous rock is illustrated in the diagram above?
        A.   felsic
        B.   intermediate
        C.   mafic
        D.   ultramafic
Answer = D (page 82)

24.   What type of rock makes up most of the Hawaiian Islands?
        A.   andesite
        B.   basalt
        C.   granite
        D.   peridotite
Answer = B (page 86)

25.   What is the effect of water on melting?
    A.   Water raises the melting temperature of a rock.
    B.   Water lowers the melting temperature of a rock.
    C.   Water raises or lowers the melting temperature of a rock depending on
           the rock composition.
    D.   Water does not change the melting temperature of a rock.
    Answer = B (pages 83–84)

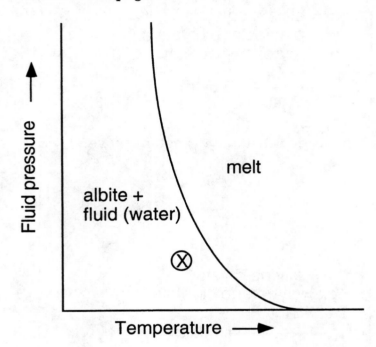

26.   Which of the following scenarios would lead to melting of a rock located at point X
       in the diagram above?
    A.   increasing the fluid pressure
    B.   increasing the temperature
    C.   either increasing the fluid pressure or increasing the temperature
    D.   neither increasing the fluid pressure nor increasing the temperature
    Answer = C (page 84)

27.   During crystallization of a magma, plagioclase feldspar _____.
    A.   is replaced by potassium feldspar
    B.   is replaced by quartz
    C.   becomes richer in calcium
    D.   becomes richer in sodium
    Answer = D (page 87)

28.   Which mineral is not part of the discontinuous reaction series?
        A.   plagioclase feldspar
        B.   olivine
        C.   amphibole
        D.   pyroxene
Answer = A (page 88)

29.   At approximately what temperature do olivine and calcium-rich plagioclase feldspar crystallize from a magma?
        A.   500°C
        B.   1000°C
        C.   2000°C
        D.   4000°C
Answer = B (page 92)

30.   What composition is the Palisades sill located near New York?
        A.   peridotite
        B.   basalt
        C.   andesite
        D.   rhyolite
Answer = B (page 90)

31.   Which of the following minerals crystallizes first from a basaltic magma?
        A.   quartz
        B.   biotite
        C.   pyroxene
        D.   olivine
Answer = D (page 92)

32.   At approximately what temperature do quartz and potassium feldspar crystallize from a magma?
        A.   600°C
        B.   1200°C
        C.   2400°C
        D.   6000°C
Answer = A (page 92)

33.   How much basaltic magma must undergo fractional crystallization in order to produce 1 cubic kilometer of granite?
        A.   0.1 cubic kilometers of basaltic magma
        B.   1 cubic kilometers of basaltic magma
        C.   10 cubic kilometers of basaltic magma
        D.   100 cubic kilometers of basaltic magma
Answer = C (page 93)

34.   How does a magma make space for itself in order to rise through the crust?
      A.   by wedging open the overlying rock
      B.   by breaking off large blocks of rock that sink into the magma chamber
      C.   by melting surrounding rocks
      D.   all of these
Answer = D (page 95)

35.   Which of the following igneous bodies is a concordant intrusive rock?
      A.   dike
      B.   pluton
      C.   sill
      D.   stock
Answer = C (page 96)

36.   How can a sill be distinguished from a lava flow?
      A.   A sill is generally coarser-grained than a lava flow.
      B.   The rocks above and below a sill will show evidence of heating, but only the rocks below a lava flow will show evidence of heating.
      C.   Sills generally do not have vesicles; lava flows generally do have vesicles.
      D.   all of these
Answer = D (pages 96–97)

37.   At what temperature do minerals crystallize in hydrothermal veins?
      A.   approximately 300°C
      B.   approximately 600°C
      C.   approximately 1000°C
      D.   approximately 2000°C
Answer = A (page 98)

38.   Which of the following is not a volcanic arc that overlies a subduction zone?
      A.   Aleutian Islands (Alaska)
      B.   Hawaiian Islands
      C.   Cascade Range (northern California, Oregon, and Washington)
      D.   Japanese Islands
Answer = B (page 99)

39.   Which of the following magma compositions will be produced by partial melting of the upper mantle?
      A.   felsic
      B.   intermediate
      C.   mafic
      D.   ultramafic
Answer = C (page 93)

40.   Which of the following is not a volcanic rock?
          A.   obsidian
          B.   pluton
          C.   pumice
          D.   tuff
       Answer = B (page 78)

41.   What type of rock would contain 10-millimeter-long plagioclase crystals "floating" in
       a fine-grained matrix of 0.5-millimeter long crystals?
          A.   pluton
          B.   porphyry
          C.   tuff
          D.   obsidian
       Answer = B (page 78)

42.   What type of silicate minerals crystallize at the highest temperature?
          A.   double chain silicates
          B.   sheet silicates
          C.   framework silicates
          D.   isolated tetrahedra silicates
       Answer = D (page 89)

43.   As a magma composition changes from mafic to felsic, which of the following ele-
       ments decreases?
          A.   sodium
          B.   potassium
          C.   iron
          D.   silicon
       Answer = C (page 82)

44.   Which of the following statements is false?
          A.   Mafic magmas are more viscous than felsic magmas.
          B.   Mafic magmas are hotter than felsic magmas.
          C.   Mafic magmas contain more calcium than felsic magmas.
          D.   Mafic magmas contain less silica than felsic magmas.
       Answer = A (page 82)

45.   Which of the following pairs of intrusive and extrusive rocks have the same chemi-
       cal composition?
          A.   granite and andesite
          B.   diorite and basalt
          C.   gabbro and basalt
          D.   gabbro and rhyolite
       Answer = C (page 81)

46.   What mineral is the most abundant in igneous rocks of intermediate composition?
         A.   quartz
         B.   pyroxene
         C.   muscovite
         D.   plagioclase feldspar
      Answer = D (page 81)

47.   Which of the following properties does not depend on the chemical composition of an igneous rock/magma?
         A.   grain size
         B.   melting temperature
         C.   mineralogy
         D.   viscosity
      Answer = A (page76)

Bowen's reaction series is depicted in the following diagram.

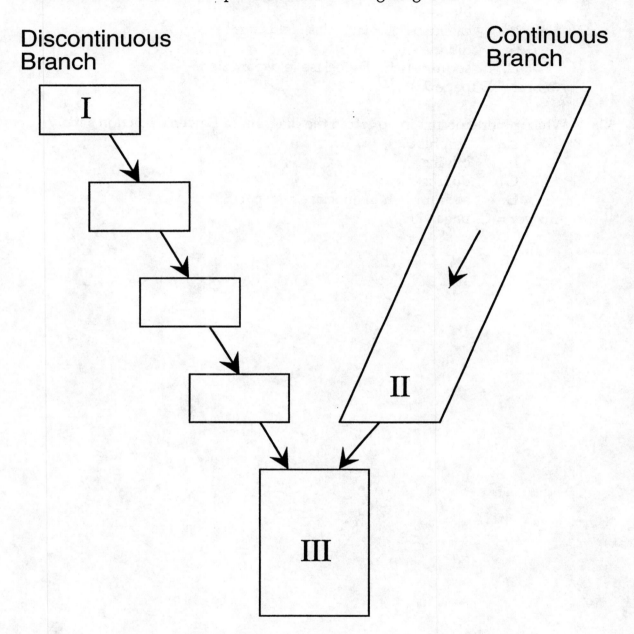

Discontinuous Branch

Continuous Branch

I

II

III

48. What mineral occurs in box I in the diagram of Bowen's reaction series?
    A. amphibole
    B. calcium-rich plagioclase feldspar
    C. olivine
    D. quartz
    Answer = C (page 92)

49.   What mineral occurs at point II in the diagram of Bowen's reaction series?
      A.   biotite
      B.   calcium-rich plagioclase feldspar
      C.   olivine
      D.   sodium-rich plagioclase feldspar
Answer = D (page 92)

50.   What mineral occurs in box III in the diagram of Bowen's reaction series?
      A.   amphibole
      B.   olivine
      C.   quartz
      D.   sodium-rich plagioclase feldspar
Answer = C (page 92)

5

Volcanism

1. Volcanic rocks cover approximately how much of the Earth's surface (including the seafloor and the land)?
    A. 20%
    B. 40%
    C. 60%
    D. 80%
    Answer = D (page 105)

2. Which of the following statements about lava is true?
    A. The viscosity of a lava increases as the silica content decreases.
    B. High temperature lavas are less viscous than low temperature lavas.
    C. The more gas a lava contains, the less violent the eruption.
    D. All of the statements above are true.
    Answer = B (page 107)

3. Which of the following types of lava tends to flow long distances?
    A. andesite
    B. basalt
    C. rhyolite
    D. dacite
    Answer = B (page 107)

4. What is the eruption temperature of basaltic lava?
    A. 400–600°C
    B. 600–800°C
    C. 800–1000°C
    D. 1000–1200°C
    Answer = D (page 107)

5. Which of the following volcanic features is not generally made up of basalt?
    A.   pahoehoe flows
    B.   pillow lavas
    C.   shield volcanoes
    D.   volcanic domes
    Answer = D (page 111)

6. Fine-grained (< 2 mm) pyroclastic material is called _____.
    A.   ash
    B.   breccia
    C.   tuff
    D.   pahoehoe
    Answer = A (page 109)

7. Which of the following statements regarding extrusive rocks is true?
    A.   Andesites contain more silica than basalts or rhyolites.
    B.   Basalts contain more silica than andesites or rhyolites.
    C.   Rhyolites contain more silica than andesites or basalts.
    D.   Andesites, basalts, and rhyolites all contain the same amount of silica.
    Answer = C (page 108)

8. What is the maximum speed of a pyroclastic flow?
    A.   20 km per hour
    B.   50 km per hour
    C.   100 km per hour
    D.   200 km per hour
    Answer = D (page 110)

9. Shield volcanoes are composed of what type of lava?
    A.   andesite
    B.   basalt
    C.   rhyolite
    D.   dacite
    Answer = B (page 111)

10. Which of the following volcanic deposits is most likely to form from a felsic lava?
    A.   pahoehoe
    B.   flood basalt
    C.   volcanic dome
    D.   shield volcano
    Answer = C (page 111)

11.   A volcano that consists of both lava flows and pyroclastic deposits is called a _____.
          A.   stratovolcano
          B.   shield volcano
          C.   cinder cone
          D.   volcanic dome
       Answer = A (page 112)

12.   Which of the following volcanoes is not a composite volcano?
          A.   Kilauea, Hawaii
          B.   Mt. St. Helens, Washington
          C.   Mt. Fujiyama, Japan
          D.   Mt. Vesuvius, Italy
       Answer = A (page 113)

13.   The eruption of vast quantities of superheated steam is called a _____.
          A.   resurgent caldera
          B.   fissure eruption
          C.   lava flow
          D.   phreatic explosion
       Answer = D (page 115)

14.   Shiprock, New Mexico, is an example of a _____.
          A.   lava flow
          B.   diatreme
          C.   cinder cone
          D.   caldera
       Answer = B (page 117)

15.   Which of the following volcanic eruptions was the smallest?
          A.   Mt. St. Helens—1980 A.D.
          B.   Long Valley caldera—700,000 years ago
          C.   Krakatoa—1883 A.D.
          D.   Yellowstone—600,000 years ago
       Answer = A (page 115)

16.   How much area is covered by the Columbia River flood basalts in the northwest United States?
          A.   200 square miles
          B.   2,000 square miles
          C.   200,000 square miles
          D.   2 million square miles
       Answer = C (page 118)

17.   Huge mudflows made up of wet volcanic debris are called _____ .
           A.   calderas
           B.   diatremes
           C.   tuffs
           D.   lahars
       Answer = D (page 119)

18.   Almost all of the 25,000 people killed in the 1985 eruption of Nevada del Ruiz in
       Columbia were killed by _____.
           A.   a large volcanic ash fall
           B.   a large lava flow
           C.   a large mudflow of volcanic debris
           D.   collapse of a volcanic caldera
       Answer = C (page 119)

19.   Which of the following gases is the main constituent of volcanic gas?
           A.   water vapor
           B.   nitrogen
           C.   carbon dioxide
           D.   sulfur dioxide
       Answer = A (page 119)

20.   What percentage of Earth's active volcanoes occur at convergent plate boundaries?
           A.   approximately 20%
           B.   approximately 50%
           C.   approximately 80%
           D.   approximately 95%
       Answer = C (page 121)

21.   Which of the following volcanic islands lies on top of a divergent plate margin?
           A.   Philippines
           B.   Hawaii
           C.   Iceland
           D.   Japan
       Answer = C (page 123)

22.   Which of the following volcanoes occurs where two tectonic plates are converging?
           A.   Fujiyama
           B.   Mt. Pinatubo
           C.   Mt. St. Helens
           D.   all of these
       Answer = D (pages 123–124)

23. In which of the following tectonic settings would you be most likely to find andesitic lavas?
      A. divergent plate boundaries
      B. transform plate boundaries
      C. ocean-ocean convergent plate boundaries
      D. ocean-continent convergent plate boundaries
   Answer = D (page 124)

24. Which of the following statements about the May 18, 1980 climactic eruption of Mt. St. Helens is false?
      A. Earthquakes began to occur underneath the volcano five years before the eruption.
      B. Volcanic ash was erupted up to 25 kilometers into the atmosphere.
      C. Approximately 60 people were killed by the eruption.
      D. The eruption was immediately preceded by a large earthquake and landslide.
   Answer = A (page 128)

25. Which of the following terms does not apply to the Hawaiian Islands?
      A. aseismic ridge
      B. hot spot
      C. shield volcanoes
      D. ash flow deposits
   Answer = D (page 125)

26. Approximately how many volcanoes erupt each year?
      A. 1
      B. 5
      C. 20
      D. 50
   Answer = D (page 126)

27. Volcanoes have killed how many people over the past 500 years?
      A. 20,000
      B. 200,000
      C. 2 million
      D. 20 million
   Answer = B (page 126)

28.   Which of the following lists is written correctly in order of increasing silica content?
        A.   rhyolite, andesite, basalt
        B.   andesite, basalt, rhyolite
        C.   basalt, andesite, rhyolite
        D.   rhyolite, basalt, andesite
Answer = C (page 107)

29.   Which of the following lava types erupts in sheets from fissures to build a lava plateau?
        A.   andesite
        B.   basalt
        C.   rhyolite
        D.   all of these
Answer = B (page 118)

30.   The largest and most violent volcanic eruptions are associated with _____.
        A.   shield volcanoes
        B.   fissure eruptions
        C.   cinder cones
        D.   resurgent calderas
Answer = D (page 114)

31.   Which of the following states does not have any active volcanoes?
        A.   Alaska
        B.   Texas
        C.   Hawaii
        D.   Washington
Answer = B (page 130)

32.   Where does most basaltic magma originate?
        A.   in the Earth's asthenosphere
        B.   in the Earth's lithosphere
        C.   in the Earth's crust
        D.   in the Earth's core
Answer = A (page 106)

33.   What is the name for basaltic lava that has a very rough, jagged surface?
        A.   aa
        B.   diatreme
        C.   lahar
        D.   pahoehoe
Answer = A (page 108)

34.   Which of the following volcanic flows moves the most rapidly?
           A.   basaltic lava flows
           B.   rhyolitic lava flows
           C.   pyroclastic flows
           D.   lahars (volcanic mudflows)
       Answer = C (page 110)

35.   The Cascade volcanoes are _____.
           A.   associated with a divergent plate margin
           B.   associated with a transform plate margin
           C.   associated with a convergent plate margin
           D.   not associated with a plate margin
       Answer = C (page 124)

36.   Most of the world's volcanoes occur along the margins of which of the following
       ocean basins?
           A.   Arctic Ocean
           B.   Atlantic Ocean
           C.   Indian Ocean
           D.   Pacific Ocean
       Answer = D (page 122)

37.   What is the name of the plate subducting beneath South America that gives rise to
       the volcanoes of the Andes?
           A.   Nazca
           B.   Pacific
           C.   Gorda
           D.   North America
       Answer = A (page 122)

38.   Approximately how many people were killed by the eruption of Mt. St. Helens?
           A.   less than 100
           B.   1,000
           C.   10,000
           D.   more than 100,000
       Answer = A (page 128)

39.   What type of volcanic material can be deposited thousands of miles away from the
       volcano?
           A.   felsic lava flows
           B.   ash flow deposits
           C.   lahar deposits
           D.   volcanic cinders
       Answer = B (page 109)

40.   Approximately how many active volcanoes are there on Earth?
        A.   50,000
        B.   5,000
        C.   500
        D.   50
        Answer = C (page 126)

41.   Solidified fragments of volcanic material ejected into the air are called _____.
        A.   phenocrysts
        B.   vesicles
        C.   pillow basalts
        D.   pyroclasts
        Answer = D (page 109)

42.   What types of lavas are erupted at convergent plate boundaries?
        A.   andesites
        B.   basalts
        C.   rhyolites
        D.   all of the above
        Answer = D (page 123)

43.   What of the following volcanoes is not part of the Cascade range?
        A.   Mt. St. Helens
        B.   Mt. Rainier
        C.   Mt. Shasta
        D.   All of these are part of the Cascade range.
        Answer = D (page 130)

44.   Which of the following is least likely to signal an impending volcanic eruption?
        A.   earthquakes
        B.   gas emissions
        C.   tsunamis
        D.   swelling of the volcano
        Answer = C (page 128)

45.   In which country are many houses heated by hot water from volcanic springs?
        A.   Iceland
        B.   Japan
        C.   Mexico
        D.   United States
        Answer = A (page 131)

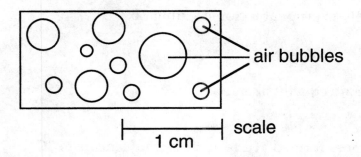

air bubbles

scale

1 cm

46. In the basalt sample depicted above, the vesicles (small spherical cavities) most likely formed by _____.
   A. the escape of gas bubbles dissolved in the lava during the eruption
   B. the weathering and erosion of olivine crystals after the eruption
   C. air bubbles entrained in the flow during the eruption
   D. vaporization of sea water during underwater eruption
   Answer = A (page 109)

47. Old Faithful geyser is associated with which of the following volcanoes?
   A. Mt. St. Helens
   B. Yellowstone
   C. Mt. Rainier
   D. Long Valley Caldera
   Answer = B (page 121)

This is a plate tectonic diagram:

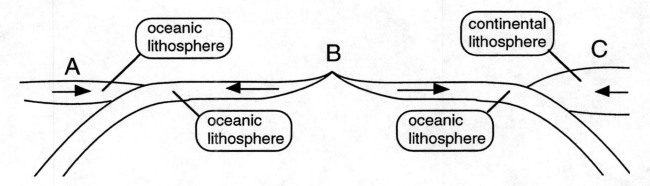

48. Which of the following volcanic chains formed at a tectonic setting similar to "A"?
   A. Aleutian Islands
   B. Andes Mountains
   C. Cascade Range
   D. Hawaiian Islands
   Answer = A (page 123)

49.   What type of lava is most likely to erupt at tectonic setting "B"?
       A.   andesitic
       B.   basaltic
       C.   rhyolitic
       D.   All of the above are equally likely.
       Answer = B (page 121)

50.   Which of the following volcanoes formed at a tectonic setting similar to "C"?
       A.   Hekla, Iceland
       B.   Mauna Loa, USA
       C.   Mount Pelee, Martinique
       D.   Mount St. Helens, USA
       Answer = D (page 124)

6

Weathering and Erosion

1. What is the term for the general process by which rocks are broken down at the Earth's surface?
   A. deposition
   B. erosion
   C. lithification
   D. weathering
   Answer = D (page 135)

2. Which of the following affect the rate of weathering?
   A. the soil
   B. the rock type
   C. the climate
   D. all of these
   Answer = D (page 138)

3. Which of the following rock types will dissolve most rapidly in a hot, humid climate?
   A. granite
   B. shale
   C. limestone
   D. all of these rocks will weather at the same rate
   Answer = C (page 136)

4. Soil is composed of _____.
   A. organic matter produced by organisms
   B. fragments of bedrock
   C. clay minerals formed by alteration of bedrock
   D. all of these
   Answer = D (page 137)

5.   In which of the following climates will chemical weathering be most rapid?
       A.   hot and dry
       B.   hot and humid
       C.   cold and dry
       D.   cold and humid
Answer = B (page 137)

6.   Which of the following statements about weathering is false?
       A.   Rocks of different compositions weather at different rates.
       B.   Heat and heavy rainfall increase the rate of chemical weathering.
       C.   The presence of soil slows down weathering of the underlying bedrock.
       D.   The longer a rock is exposed at the surface, the more weathered it becomes.
Answer = C (pages 138–139)

7.   Which of the following minerals has a low solubility and therefore is least susceptible to chemical weathering?
       A.   calcite
       B.   plagioclase feldspar
       C.   quartz
       D.   pyroxene
Answer = C (page 138)

8.   Which of the following chemical species is not produced by the chemical weathering of potassium feldspar (orthoclase)?
       A. $Al_2Si_2O_5(OH)_4$ (kaolinite, a clay mineral)
       B. H_2O (water)
       C. K^+ (potassium ion)
       D. SiO_2 (dissolved silica)
Answer = B (page 142)

9.   Chemical weathering is sped up by acids. The most common natural acid on the Earth's surface is _____.
       A.   nitric acid
       B.   hydrochloric acid
       C.   carbonic acid
       D.   sulfuric acid
Answer = C (page 142)

10.   Of the following gases, which is the least abundant in the Earth's atmosphere?
        A.   argon
        B.   oxygen
        C.   carbon dioxide
        D.   nitrogen
        Answer = C (page 142)

11.   Carbon dioxide makes up _____ of the Earth's atmosphere.
        A.   less than 0.1 percent
        B.   approximately 1 percent
        C.   approximately 10 percent
        D.   approximately 50 percent
        Answer = A (page 142)

12.   What of the following acids is not an abundant acid in acid rain?
        A.   nitric acid
        B.   hydrochloric acid
        C.   carbonic acid
        D.   sulfuric acid
        Answer = B (page 142)

13.   Carbonic acid, the primary agent of chemical weathering, is produced by _____.
        A.   carbon dioxide dissolving in rainwater
        B.   plant roots
        C.   bacteria that feed on plant and animal remains
        D.   all of these
        Answer = D (page 143)

14.   Which of the following minerals is least likely to form a clay mineral during weathering?
        A.   feldspar
        B.   quartz
        C.   amphibole
        D.   mica
        Answer = B (page 144)

15.   The major source of aluminum metal, _____, is a clay-rich ore composed of aluminum hydroxide.
        A.   hematite
        B.   bauxite
        C.   kaolinite
        D.   montmorillonite
        Answer = B (page 144)

16.   The symbol, Fe^{3+}, refers to _____.
        A.   iron metal
        B.   ferrous iron
        C.   ferric iron
        D.   hematite
Answer = C (page 144)

17.   Caves are most common in which of the following rock types?
        A.   granite
        B.   limestone
        C.   basalt
        D.   sandstone
Answer = B (page 146)

18.   Which of the following minerals is least stable at the Earth's surface?
        A.   calcite
        B.   halite
        C.   olivine
        D.   quartz
Answer = C (page 147)

19.   Which of the following minerals is most stable at the Earth's surface?
        A.   hematite
        B.   mica
        C.   olivine
        D.   pyroxene
Answer = A (page 147)

20.   Which of the following processes is not an example of chemical weathering?
        A.   dissolution of calcite
        B.   breakdown of feldspar to form kaolinite
        C.   splitting of a rock along a fracture
        D.   rusting of a nail
Answer = C (page 148)

21.   As a rock breaks into smaller pieces, the surface area to volume ratio _____.
        A.   increases
        B.   decreases
        C.   remains the same
        D.   can increase or decrease depending on the size of the pieces
Answer = A (page 141)

22.   A substance that releases hydrogen ions (H⁺) to a solution is called a(n) _____.
         A.   oxidizer
         B.   pedalfer
         C.   acid
         D.   hydrater
      Answer = C (page 142)

23.   What is the term geologists use for the layer of loose, heterogeneous weathered material lying on top of the bedrock?
         A.   humus
         B.   laterite
         C.   regolith
         D.   soil
      Answer = C (page 152)

24.   The uppermost layer of soil is called the _____.
         A.   A-horizon
         B.   B-horizon
         C.   C-horizon
         D.   regolith
      Answer = A (page 154)

25.   The characteristics of a soil depend on _____.
         A.   the climate
         B.   the type of bedrock
         C.   the length of time the soil has had to develop
         D.   all of these
      Answer = D (page 154)

26.   Which of the following farming practices helps prevent the erosion of topsoil?
         A.   plowing a field perpendicular to contour lines
         B.   plowing a field parallel to contour lines
         C.   plowing a field in the direction that water drains
         D.   *all* of these will help prevent the erosion of top soil
      Answer = B (pages 152–153)

27.   Which of the following soil types has large amounts of organic matter in the A-horizon?
         A.   aridosol
         B.   entisol
         C.   mollisol
         D.   vertisol
      Answer = C (page 156)

28.   What is the name for soils that are rich in calcium?
    A.   laterites
    B.   evaporites
    C.   pedocals
    D.   pedalfers
    Answer = C (page 158)

29.   Which of the following types of soil is the most fertile?
    A.   laterites
    B.   evaporites
    C.   pedocals
    D.   pedalfers
    Answer = D (page 158)

30.   The physical and chemical weathering of a granite will produce _____.
    A.   ions dissolved in rainwater and soil water
    B.   mineral fragments and granite fragments
    C.   clay minerals and iron oxides
    D.   all of these
    Answer = D (page 159)

31.   Which of the following is not a major soil type?
    A.   pedalfer
    B.   laterite
    C.   kaolinite
    D.   pedocal
    Answer = C (page 154)

32.   In a soil profile, organic matter is found _____.
    A.   only in the A-horizon
    B.   mostly in the A-horizon with some in the B-horizon
    C.   mostly in the A-horizon with some in the B- and C-horizons
    D.   in the A-, B-, and C-horizons in equal amounts
    Answer = B (page 154)

33.   What type of soils form in a hot, dry climate like the southwestern U.S.?
    A.   laterites
    B.   evaporites
    C.   pedocals
    D.   pedalfers
    Answer = C (page 158)

34. In the atmosphere, carbonic acid forms from the reaction of carbon dioxide and
_____.
    A. fossil fuels
    B. nitrogen
    C. oxygen
    D. water
Answer = D (page 142)

35. What happens to the potassium (K) in feldspar during chemical weathering?
    A. It dissolves in the water.
    B. It becomes part of the clay mineral.
    C. It evaporates.
    D. It becomes concentrated in potassium metal deposits.
Answer = A (page 142)

36. What is the difference between ferrous iron and ferric iron?
    A. Ferrous iron contains less electrons than ferric iron.
    B. Ferrous iron contains more electrons than ferric iron.
    C. Ferrous iron contains less neutrons than ferric iron.
    D. Ferrous iron contains more neutrons than ferric iron.
Answer = B (page 144)

37. Which of the following minerals found in a granite is not altered by chemical weathering?
    A. biotite
    B. feldspar
    C. magnetite
    D. quartz
Answer = D (page 140)

38. The deep red color of soils found in Georgia and other warm, humid regions is
caused by _____.
    A. iron oxides
    B. clay minerals
    C. quartz
    D. feldspar
Answer = A (page 145)

39.   Which of the following minerals would be most likely to form a clay mineral during weathering?
          A.   iron oxide
          B.   mica
          C.   calcite
          D.   quartz
       Answer = B (page 144)

40.   Which of the following silicates is the least stable at the Earth's surface?
          A.   quartz
          B.   biotite
          C.   olivine
          D.   amphibole
       Answer = C (page 147)

41.   Which of the following is a clay mineral?
          A.   amphibole
          B.   calcite
          C.   kaolinite
          D.   muscovite
       Answer = C (page 141)

42.   Which of the following is *not* a product of chemical weathering?
          A.   plagioclase feldspar
          B.   dissolved silica
          C.   iron oxides
          D.   clay minerals
       Answer = A (page 143)

WEATHERING RATE

slow ⟶ ⟶ fast

43.   Which of the following factors would increase the weathering rate as depicted in the diagram above?
          A.   increasing rainfall
          B.   increasing temperature
          C.   increasing organic activity
          D.   *all* of these
       Answer = D (page 138)

WEATHERING RATE

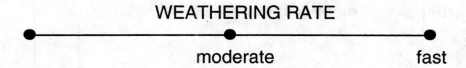

moderate   fast

44.   Which of the following conditions promotes slow weathering?
       A.   cold temperatures
       B.   thick soils
       C.   high rainfall
       D.   fracturing in rocks
       Answer = A (page 138)

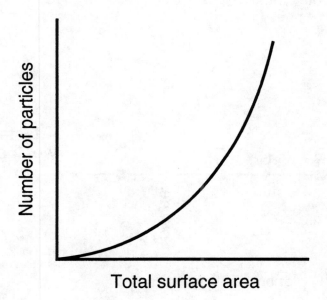

45.   The diagram above depicts how the total surface area of a rock mass varies with the number of particles *for a fixed volume of rock*. Based on this diagram, what can be said about the ratio of surface area to volume?
       A.   The ratio decreases as the number of particles increases.
       B.   The ratio increases as the number of particles increases.
       C.   The ratio first decreases then increases as the number of particles increases.
       D.   The ratio does not depend on the number of particles.
       Answer = B (page 141)

Relative weathering rates of different minerals.

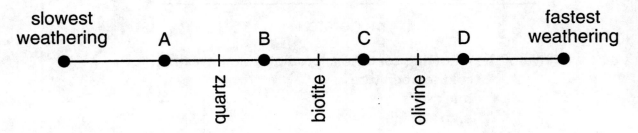

46.   Where would the mineral amphibole on the diagram?
   A. point A
   B. point B
   C. point C
   D. point D
Answer = C (page 147)

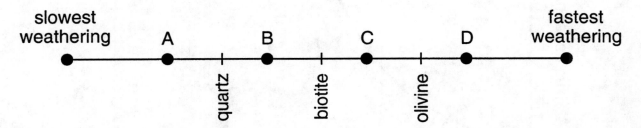

47.   Where would hematite (iron oxide) plot on the diagram?
   A. point A
   B. point B
   C. point C
   D. point D
Answer = A (page 147)

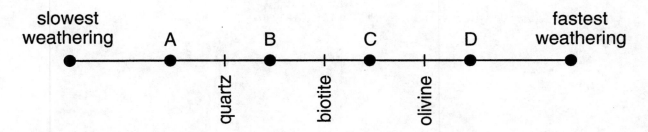

48. Where would limestone plot on the diagram?
    A. point A
    B. point B
    C. point C
    D. point D
    Answer = D (page 147)

49. What process resulted in the unusual typography of Half Dome in Yosemite National Park in California?
    A. chemical weathering
    B. dissolution
    C. exfoliation
    D. spheroidal weathering
    Answer = C (page 150)

50. Which of the following soil types is characteristic of Tempe, Arizona?
    [replace "Tempe, Arizona" with the location of your college/university]
    A. aridisols
    B. mollisols
    C. oxisols
    D. ultisols
    Answer = A for Tempe, Arizona (pages 156–157)

51. Which of the following human activities has resulted in increased rates of weathering?
    A. The release of sulfur and nitrogen oxides that cause acid rain
    B. The physical disintegration of rocks during construction and mining
    C. Both A and B
    D. Neither A nor B
    Answer = C (page 158)

7

Sediments and Sedimentary Rocks

1.   Which of the following lists is written in order of increasing temperature?
       A.   sedimentation, metamorphism, diagenesis
       B.   diagenesis, sedimentation, metamorphism
       C.   sedimentation, diagenesis, metamorphism
       D.   metamorphism, diagenesis, sedimentation
    Answer = C (page 166)

2.   Which of the following processes does not occur during diagenesis?
       A.   compaction
       B.   cementation
       C.   lithification
       D.   metamorphism
    Answer = D (page 166)

3.   Which of the following types of sediments are most abundant?
       A.   biochemical sediments
       B.   chemical sediments
       C.   clastic sediments
       D.   all of these sediments occur in approximately equal amounts
    Answer = C (page 167)

4.   The tendency for variations in current velocity to segregate sediments on the basis of
      particle size is called _____.
            A.   lithification
            B.   compaction
            C.   metamorphism
            D.   sorting
      Answer = D (page 168)

5.   How fast are moderate strength currents in most rivers that carry and deposit sand
      in their channels?
            A.   2 to 5 cm/s
            B.   5 to 20 cm/s
            C.   20 to 50 cm/s
            D.   50 to 200 cm/s
      Answer = C (page 168)

6.   Which of the following types of currents can transport sand grains?
            A.   rivers
            B.   wind
            C.   ocean waves
            D.   all of these
      Answer = D (pages 169–170)

7.   With time, sediment transported by glaciers _____.
            A.   tends to become rounded
            B.   tends to become smaller
            C.   tends to become rounded and smaller
            D.   none of these
      Answer = B (page 170)

8.   Which of the sand grains depicted above has been transported the furthest?
            A.   grain 1
            B.   grain 2
            C.   grain 3
            D.   grain 4
      Answer = C (page 170)

9.   In which of the following sedimentary environments would you expect the sand deposits to be poorly sorted?
     A.   alluvial
     B.   beach
     C.   desert
     D.   glacial
Answer = D (page 170)

10.   Coal is formed from _____.
     A.   gas
     B.   oil
     C.   peat
     D.   limestone
Answer = C (page 171)

11.   Which of the following terms describes a set of sedimentary environments that exist at the same time in different parts of a region?
     A.   bedding sequence
     B.   clastic sediments
     C.   diagenesis
     D.   sedimentary facies
Answer = D (page 173)

12.   Which of the following environments is an example of a shoreline environment?
     A.   alluvial
     B.   continental shelf
     C.   deltaic
     D.   organic reef
Answer = C (page 173)

13.   In which of the following sedimentary environments would you least expect to find gravel?
     A.   alluvial
     B.   beach
     C.   deep sea
     D.   glacial
Answer = C (page 174)

14.   Which of the following sedimentary environments is characterized by sand, gravel, and mud?
      A.   deep sea
      B.   evaporite
      C.   glacial
      D.   swamp
Answer = C (page 174)

15.   Siliceous environments, named for the silica shells deposited in them, occur _____.
      A.   in an evaporite environment
      B.   in a swamp environment
      C.   in a reef environment
      D.   in a deep-sea environment
Answer = D (page 174)

16.   What type(s) of clastic sediments are deposited in the deep sea?
      A.   mud
      B.   sand
      C.   mud and sand
      D.   mud, sand, and gravel
Answer = A (page 174)

17.   Which of the following sedimentary environments is dominated by waves and tidal currents?
      A.   alluvial
      B.   deltaic
      C.   deep sea
      D.   desert
Answer = B (page 174)

Cross section of a sand dune.

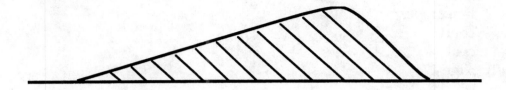

18.   Assuming the sand dune was deposited by wind currents, which way is the wind blowing?

A.

B.

C. ◄────────►

D.   Wind currects cannot be determined from the information given.
Answer = B (page 175)

Cross section of a sand dune.

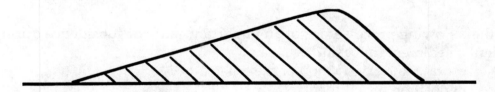

19.   The diagonal layers are called _____.
        A.   cross beds
        B.   graded beds
        C.   point bars
        D.   ripples
Answer = A (page 175)

20.   Most shells of marine organisms are composed of _____.
        A.   silica
        B.   calcium carbonate
        C.   calcium phosphate
        D.   calcium sulfate
Answer = B (page 175)

21.   What is the most abundant biochemical precipitate in the oceans?
      A.   halite (rock salt)
      B.   limestone
      C.   chert
      D.   coal
    Answer = B (page 175)

22.   Ripples occur _____.
      A.   on sand beneath the waves at beaches
      B.   on underwater sandbars in streams
      C.   on the surface of windswept dunes
      D.   in all of these environments
    Answer = D (page 176)

23.   In which of the following sedimentary environments would you most likely find symmetrical ripples?
      A.   alluvial
      B.   beach
      C.   deep-sea
      D.   desert
    Answer = B (page 176)

24.   Which of the following processes is not an important cause of subsidence during the development a sedimentary basin?
      A.   cooling and contraction of the crust
      B.   deposition of sediments
      C.   erosion of sediments
      D.   tectonic downfaulting
    Answer = C (pages 178–179)

25.   What is the approximate temperature of a sediment that is buried to a depth of 3 km?
      A.   0°C
      B.   100°C
      C.   300°C
      D.   1000°C
    Answer = B (page 180)

26.   Which of the following processes occurs during lithification?
      A.   recrystallization of unstable minerals
      B.   compaction
      C.   cementation
      D.   All of these may occur during lithification.
    Answer = D (page 180)

27.   Which of the following is an example of a physical, as opposed to chemical, diagenetic process?
        A.   cementation
        B.   compaction
        C.   dissolution
        D.   All of the above are physical diagenetic processes.
Answer = B (page 180)

28.   What is the porosity of newly deposited mud?
        A.   less than 5%
        B.   between 5% and 20%
        C.   between 20% and 50%
        D.   greater than 50%
Answer = D (page 180)

29.   Which of the following is an example of a clastic sedimentary rock?
        A.   chert
        B.   conglomerate
        C.   dolostone
        D.   evaporite
Answer = B (page 181)

30.   Which of the following lists is written in order of decreasing particle size?
        A.   sandstone, siltstone, conglomerate
        B.   sandstone, conglomerate, siltstone
        C.   conglomerate, sandstone, siltstone
        D.   siltstone, sandstone, conglomerate
Answer = C (page 181)

31.   Which of the following groups represent the least abundant sedimentary rocks?
        A.   limestones and dolostones
        B.   sandstones and conglomerates
        C.   cherts and evaporites
        D.   siltstones, mudstones, and shales
Answer = C (page 182)

32.   What is the difference between a breccia and a conglomerate?
        A.   Breccias are coarse grained; conglomerates are fine grained.
        B.   Conglomerates are coarse grained; breccias are fine grained.
        C.   Breccias have rounded rock fragments; conglomerates have angular rock fragments.
        D.   Conglomerates have rounded rock fragments; breccias have angular rock fragments.
Answer = D (page 182)

33.   A feldspar-rich sandstone is called a(n) _____.
         A.   arkose
         B.   graywacke
         C.   quartz arenite
         D.   shale
       Answer = A (page 184)

34.   Which of the following types of sandstones is most likely to form by the rapid
       mechanical weathering of a granite?
         A.   arkose
         B.   graywacke
         C.   quartz arenite
         D.   shale
       Answer = A (page 184)

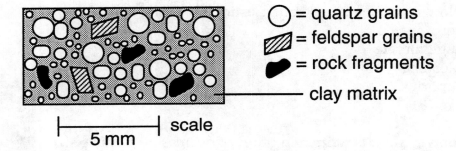

35.   What type of sandstone is depicted in the illustration above?
         A.   arkose
         B.   graywacke
         C.   lithic sandstone
         D.   quartz arenite
       Answer = B (page 184)

36.   Which of the following elements is least abundant in seawater?
         A.   calcium
         B.   magnesium
         C.   silicon
         D.   sodium
       Answer = C (page 185)

37.   The most abundant chemical/biochemical sedimentary rocks are _____.
         A.   carbonates
         B.   cherts
         C.   sandstones
         D.   shales
       Answer = A (page 185)

38.   Which of the following ions is not abundant in seawater?
      A. Ca^{2+}
      B. Fe^{2+}
      C. Mg^{2+}
      D. Na^+
Answer = B (page 185)

39.   Which of the following common sedimentary minerals does not contain calcium?
      A.   gypsum
      B.   dolomite
      C.   halite
      D.   calcite
Answer = C (page 186)

40.   Which of the following is not a clastic sedimentary environment?
      A.   alluvial
      B.   evaporite
      C.   beach
      D.   delta
Answer = B (page 186)

41.   The Bahama banks are an example of a(n) _____.
      A.   alluvial environment
      B.   carbonate platform
      C.   deltaic deposit
      D.   rift valley
Answer = B (page 187)

42.   Which of the following minerals does not precipitate directly from sea water?
      A.   halite
      B.   dolomite
      C.   calcite
      D.   gypsum
Answer = B (page 188)

43.   What is the chemical formula for dolomite?
      A. $CaMg(CO_3)_2$
      B. $CaSO_4 \cdot 2H_2O$
      C. $CaCO_3$
      D. $CaSO_4$
Answer = A (page 188)

44.   The conversion of limestone to dolostone involves the replacement of calcium ions
        with _____.
                A.   carbonate ions
                B.   magnesium ions
                C.   silica ions
                D.   sodium ions
        Answer = B (page 188)

45.   Which of the following rocks is composed of calcium carbonate?
                A.   chert
                B.   dolostone
                C.   limestone
                D.   sandstone
        Answer = C (page 188)

46.   Which of the following is deposited only by non-biological, chemical precipitation?
                A.   halite (rock salt)
                B.   limestone
                C.   chert
                D.   coal
        Answer = A (page 189)

47.   A sedimentary rock made up of biochemically-precipitated silica is called a(n)
        _____.
                A.   limestone
                B.   quartz arenite
                C.   chert
                D.   evaporite
        Answer = C (page 190)

48.   Which of the following sedimentary rocks is the product of diagenesis and has no
        exact equivalent as a sediment?
                A.   limestone
                B.   chert
                C.   phosphorite
                D.   sandstone
        Answer = C (page 190)

49.   Which of the following sedimentary rocks can form by biochemical precipitation?
                A.   chert
                B.   dolostone
                C.   gypsum
                D.   sandstone
        Answer = A (page 190)

50.   In which of the following environments does coal form?
          A.   alluvial
          B.   deltaic
          C.   evaporite
          D.   swamp
    Answer = D (page 191)

8

Metamorphic Rocks

1. Which of the following processes will cause metamorphism?
   A. a change in the chemical environment
   B. an increase in temperature
   C. an increase in pressure
   D. all of these
   Answer = D (page 195)

2. Marble is a metamorphic rock that forms from _____?
   A. granite
   B. limestone
   C. sandstone
   D. shale
   Answer = B (page 196)

3. Silicate minerals that are diagnostic of metamorphism include _____ .
   A. andalusite and calcite
   B. kyanite and quartz
   C. sillimanite and epidote
   D. feldspar and muscovite
   Answer = C (page 196)

4.   The pressure and heat that drive metamorphism result from which three forces?
     A.   The internal heat of the Earth, the weight of overlying rock, and horizontal pressures developed as rocks become deformed.
     B.   The weight of overlying rock, solar heat, and nuclear fusion
     C.   Horizontal pressures developed as rocks become deformed, covalent bonding, and heat released during rock crystallization.
     D.   The internal heat of the Earth, nuclear bonding, and heat released during chemical weathering.
     Answer = A (page 197).

5.   What is the pressure on a rock at 15 kilometers depth?
     A.   Approximately 4 times atmospheric pressure.
     B.   Approximately 40 times atmospheric pressure.
     C.   Approximately 400 times atmospheric pressure.
     D.   Approximately 4000 times atmospheric pressure.
     Answer = D (page 197)

6.   _____ is the process where rocks previously metamorphosed under high-grade conditions are later metamorphosed under low-grade conditions.
     A.   Metasomatism
     B.   Cataclasis
     C.   Foliation
     D.   Retrograde metamorphism
     Answer = D (page 197)

7.   A geothermometer is _____.
     A.   a device that measures temperature when lowered into deep drill holes
     B.   a device that measures current rock temperatures
     C.   a mineral assemblage that can reveal the maximum temperature attained by a rock
     D.   the range of temperatures experienced by a rock in its geologic history
     Answer = C (page 198)

8.   Which of the following types of pressure will result the alignment of metamorphic minerals?
     A.   contact pressure
     B.   directed pressure
     C.   confining pressure
     D.   chemical pressure
     Answer = B (page 198)

9.   Metasomatism is _____.
              A.   the change in the bulk composition of a rock during metamorphism
              B.   metamorphism caused by nearby intrusions
              C.   metamorphism caused by tectonic movements along faults
              D.   the parallel alignment of minerals in a metamorphic rock
        Answer = A (page 198)

10.   Common elements in hydrothermal fluids are _____.
              A.   sodium, potassium, and silicon
              B.   zinc, iron, and titanium
              C.   sodium, calcium, and magnesium
              D.   silicon, iron, and sulfur
        Answer = A (page 198)

11.   During metamorphism, changes in the bulk composition of a rock occur primarily as a result of _____.
              A.   increases in temperature
              B.   increases in pressure
              C.   reaction with hydrothermal fluids
              D.   all of these
        Answer = C (page 198)

12.   What type of metamorphism is caused by high temperature and high pressure imposed over a large volume of crust?
              A.   burial
              B.   contact
              C.   regional
              D.   cataclastic
        Answer = C (page 199)

13.   Hydrothermal metamorphism is very common in which of the following settings?
              A.   at continental collision zones
              B.   along shallow faults
              C.   at mid-ocean ridges
              D.   in mid-continental regions
        Answer = C (page 200)

14.   Metamorphism occurs _____.
              A.   adjacent to igneous intrusions
              B.   along faults
              C.   in subduction zones
              D.   in all of these settings
        Answer = D (page 200)

15.   A rock that has undergone cataclastic metamorphism would most likely display which of the following?
       A.   preserved sedimentary layering
       B.   pulverized rock fragments
       C.   well-defined porphyroblasts
       D.   large olivine crystals
       Answer = B (page 200)

16.   In which of the following metamorphic environments would you expect bedding and other sedimentary structures to be preserved?
       A.   burial metamorphism
       B.   regional metamorphism
       C.   cataclastic metamorphism
       D.   none of these
       Answer = A (page 200)

17.   What is the most prominent textural feature of regional metamorphic rocks?
       A.   foliation
       B.   bedding
       C.   cataclasis
       D.   ripples
       Answer = A (page 201)

18.   The parallel alignment of mica in a metamorphic rock is an example of _____.
       A.   porphyroblasts
       B.   bedding
       C.   metasomatism
       D.   foliation
       Answer = D (page 201)

19.   What is the relation between metamorphic foliation and sedimentary bedding?
       A.   Sedimentary bedding is required in order for a rock to develop a metamorphic foliation.
       B.   Sedimentary bedding and metamorphic foliation are two terms for the same phenomenon.
       C.   Sedimentary bedding and metamorphic foliation are generally parallel.
       D.   There is no regular relationship between sedimentary bedding and metamorphic foliation.
       Answer = D (page 201)

20.   Which of the following is not used to classify foliated rocks?
         A.   the crystal size
         B.   the texture of the parent rock
         C.   the degree to which minerals are segregated into lighter and darker bands
         D.   the metamorphic grade
    Answer = B (page 202)

21.   Which of the following sets is arranged in order of increasing metamorphic grade?
         A.   shale --> slate --> phyllite
         B.   phyllite --> gneiss --> schist
         C.   phyllite --> slate --> schist
         D.   schist --> shale --> gneiss
    Answer = A (page 202)

22.   Which of the following statements about the metamorphism of a shale is false?
         A.   With increasing metamorphism, the clay minerals break down to form micas.
         B.   With increasing metamorphism, the grain size of the rock gets smaller.
         C.   With increasing metamorphism, foliation develops.
         D.   With increasing metamorphism, the amount of water in the rock decreases.
    Answer = B (page 203)

23.   Which mineral is responsible for the strong foliation in a schist?
         A.   quartz
         B.   mica
         C.   calcite
         D.   pyroxene
    Answer = B (page 203)

24.   Which statement regarding hornfels is not true?
         A.   A hornfels is a high-temperature rock.
         B.   A hornfels has a granular texture.
         C.   A hornfels is defined by its mineralogy: quartz, pyroxene, and olivine.
         D.   A hornfels has randomly-oriented crystals.
    Answer = C (page 204)

25.   Which of the following metamorphic rocks can not form from a shale?
         A.   schist
         B.   marble
         C.   hornfels
         D.   slate
    Answer = B (page 204)

26.   Which of the following rocks represents the highest metamorphic grade?
          A.   slate
          B.   schist
          C.   phyllite
          D.   gneiss
        Answer = D (page 204)

27.   Light-colored rocks with coarse bands of segregated light and dark minerals are called _____.
          A.   quartzites
          B.   schists
          C.   gneisses
          D.   granulites
        Answer = C (page 204)

28.   During metamorphism, a quartz-rich sandstone will change into what type of rock?
          A.   slate
          B.   schist
          C.   quartzite
          D.   all of these
        Answer = C (page 204)

29.   If an argillite is broken, it will break along _____.
          A.   irregular or conchoidal fractures
          B.   smooth, parallel foliation planes
          C.   rough, non-parallel foliation planes
          D.   ninety degree angles representing two cleavage planes
        Answer = A (page 204)

30.   In a metamorphic rock, the larger crystals set in a finer-grained matrix are called _____.
          A.   phenocrysts
          B.   vesicles
          C.   porphyroblasts
          D.   mylonites
        Answer = C (page 205)

31.   Which of the following metamorphic minerals commonly forms porphyroblasts?
          A.   garnet
          B.   chlorite
          C.   calcite
          D.   amphibole
        Answer = A (page 205)

32.   How can a geologist recognize a fault zone that formed deep within the crust?
          A.   by the presence of mylonites
          B.   by the presence of porphyroblasts
          C.   by the presence of granulites
          D.   by the presence of shales
        Answer = A (page 205)

33.   Which of the following metamorphic rocks is not paired with its true parent rock?
          A.   greenstone, basalt
          B.   quartzite, quartz sandstone
          C.   schist, shale
          D.   hornfels, dolomite
        Answer = D (page 206)

34.   A normal geothermal gradient is approximately _____.
          A.   3,000°C/km
          B.   300°C/km
          C.   30°C/km
          D.   3°C/km
        Answer = C (page 207)

35.   Which of the following index minerals forms at the highest metamorphic grade?
          A.   chlorite
          B.   sillimanite
          C.   biotite
          D.   garnet
        Answer = B (page 207)

This is a Barrovian sequence.

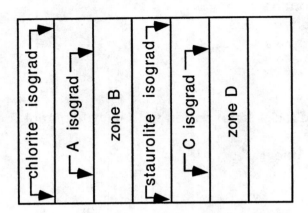

36.   In the figure above, the A isograd is the _____ isograd.
   A.   amphibole
   B.   muscovite
   C.   biotite
   D.   garnet
Answer = C (page 207)

37.   In the figure above, what mineral would not be present in zone B?
   A.   garnet
   B.   chlorite
   C.   biotite
   D.   quartz
Answer = A (page 207)

38.   In the figure above, the C isograd is the _____ isograd.
   A.   kyanite
   B.   muscovite
   C.   staurolite
   D.   sillimanite
Answer = A (page 207)

39.   In the figure above, zone D should contain which of the following minerals?
   A.   amphibole
   B.   sillimanite
   C.   calcite
   D.   kyanite
Answer = D (page 207)

40.   A rock rich in garnet and pyroxene that forms at extremely high pressures and moderate to high temperatures is called a(n) _____.
    A.   amphibolite
    B.   hornfels
    C.   granulite
    D.   eclogite
Answer = D (page 208)

41.   Greenschist, granulite, and eclogite are names of _____.
    A.   geothermometers
    B.   rocks containing good foliation
    C.   metamorphic facies
    D.   contact metamorphosed rocks
Answer = C (page 208)

42.   In which of the following metamorphic facies would you expect to see evidence of partial melting?
    A.   zeolite
    B.   blueschist
    C.   granulite
    D.   greenschist
Answer = C (page 208)

43.   Which metamorphic facies represents the lowest metamorphic pressures and temperatures?
    A.   granulite
    B.   hornfels
    C.   greenschist
    D.   zeolite
Answer = D (page 208)

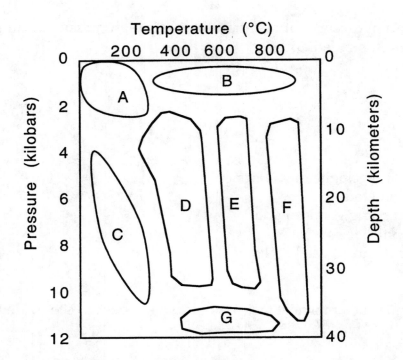

44.   The zeolite facies occurs in which pressure-temperature regime?
           A.   area A
           B.   area B
           C.   area C
           D.   area D
       Answer = A (page 208)

45.   The eclogite facies occurs in which pressure-temperature regime?
           A.   area A
           B.   area C
           C.   area E
           D.   area G
       Answer = D (page 208)

46.   The greenschist facies occurs in which pressure-temperature regime?
           A.   area C
           B.   area D
           C.   area E
           D.   area F
       Answer = B (page 208)

47. Which of the following rocks can be considered gradational between an igneous and a metamorphic rock?
    A. gneiss
    B. zeolite
    C. amphibolite
    D. migmatite
    Answer = D (page 209)

48. Which of the following scenarios would result in the largest contact aureole?
    A. intrusion of a basaltic dike (10 m wide)
    B. intrusion of a gabbroic pluton (5 km in diameter)
    C. intrusion of a rhyolitic dike (10 m wide)
    D. intrusion of a granitic pluton (5 km in diameter)
    Answer = B (page 209)

49. The reaction:
    calcite ($CaCO_3$) + quartz (SiO_2) ---> wollastonite ($CaSiO_3$) + carbon dioxide (CO_2) is an example of a _____ reaction.
    A. weathering
    B. precipitation
    C. metamorphic
    D. melting
    Answer = C (page 210)

50. Which of the following metamorphic rocks forms in the forearc of a subduction zone?
    A. amphibolite
    B. blueschist
    C. hornfels
    D. granulite
    Answer = B (page 211)

9

The Rock Record
and the Geologic Time Scale

1.   Radiometric age is often referred to as _____ age.
        A.   total
        B.   absolute
        C.   historical
        D.   geologic
    Answer = B (page 218)

2.   The principal of original horizontality states that _____.
        A.   most rocks in Earth's crust are layered horizontally
        B.   igneous intrusions form essentially horizontal layers
        C.   metamorphic gradients are essentially horizontal before deformation
        D.   sediments are deposited as essentially horizontal layers
    Answer = D (page 219)

3.   An undeformed sedimentary layer is _____ than the layer above and _____ than the layer below.
        A.   younger ... younger
        B.   younger ... older
        C.   older ... younger
        D.   older ... older
    Answer = C (page 219)

4.   A stratigraphic sequence is a vertical set of strata _____.
        A.   used as a chronological record of the geologic history of a region
        B.   that is unique to a specific area
        C.   that represents a repeating set of events such as recurring floods, debris
              flows, etc.
        D.   bounded above and below by igneous and/or metamorphic rocks
      Answer = A (page 220)

5.   We cannot accurately measure time in years with stratigraphy for several reasons.
      Which of the reasons listed below is false?
        A.   Sediments accumulate at different rates in different sedimentary envi-
              ronments.
        B.   Many sedimentary rocks contain fossils which have not been identified,
              making dating difficult.
        C.   The rock record does not tell us how many years have passed between
              periods of deposition.
        D.   Stratigraphy alone cannot be used to determine the relative ages of two
              widely separated beds.
      Answer = B (page 220)

6.   Which of the following will not make a fossil?
        A.   decomposed organic matter
        B.   plant impressions (casts)
        C.   animal footprints
        D.   loose animal bones
      Answer = A (page 221)

7.   The most common fossils in rocks of the last 500 million years are _____.
        A.   vertebrate bones
        B.   vertebrate teeth
        C.   invertebrate shells
        D.   leaves
      Answer = C (page 221)

8.   Fossils are most common in which rock types?
        A.   sedimentary
        B.   igneous
        C.   metamorphic
        D.   sedimentary, igneous, and metamorphic all commonly contain fossils
      Answer = A (page 221)

9.   Which of the following scientists did not study fossils?
        A.   Leonardo da Vinci
        B.   Nicolaus Steno
        C.   Charles Darwin
        D.   Henri Becquerel
Answer = D (page 222)

10.   The study of faunal succession allows _____.
        A.   matching of similarly-aged rocks from different outcrops
        B.   absolute dating of fossil-bearing strata
        C.   correlation of fossils with mammals of today
        D.   the study of habits of extinct mammals
Answer = A (page 222)

11.   Who is credited with developing the principle of faunal succession?
        A.   Charles Lyell
        B.   Charles Darwin
        C.   William Smith
        D.   Nicolaus Steno
Answer = C (page 222)

12.   A disconformity is _____.
        A.   a rock unit that does not contain fossils
        B.   an erosional surface between igneous and metamorphic rocks
        C.   an erosional surface between horizontal layers of sedimentary rocks
        D.   a rock unit that is different than units above or below it
Answer = C (page 223)

13.   A nonconformity is _____.
        A.   a rock unit that is different than units above or below it
        B.   a gap in the geologic record bounded below by metamorphic or igneous rocks and bounded above by sedimentary rocks
        C.   a rock unit that does not contain fossils
        D.   a sequence of rocks that does not contain any gaps in the geologic record
Answer = B (page 223)

14.   What is the name for an erosion surface that separates two sets of sedimentary layers with non-parallel bedding planes?
        A.   cross-bedding
        B.   formation
        C.   angular unconformity
        D.   fault
Answer = C (page 223)

15.   Sequence stratigraphy was previously known as _____.
      A.   faunal stratigraphy
      B.   horizontal stratigraphy
      C.   seismic stratigraphy
      D.   characteristic stratigraphy
   Answer = C (page 224)

16.   Seismic stratigraphy was so named because _____.
      A.   exploration seismology allowed geologists to see individual beds on a
            seismic profile or cross-section
      B.   seismic charges were used to blast holes and geologists mapped the
            units in the blast holes
      C.   seismic charges were used during road construction and geologists
            mapped the rocks exposed in the road cuts
      D.   seismology detected earthquake focal points and geologists mapped
            the area around the focal points.
   Answer = A (page 225)

In the following geologic cross section, units A, B, C, D, E, and F are sedimentary
rocks. Unit G is a granite.

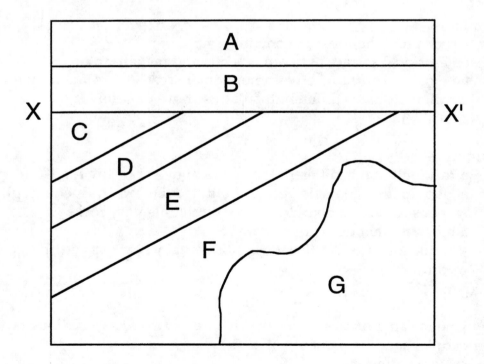

17. Which of the following units is the youngest?
    A. unit A
    B. unit B
    C. unit C
    D. unit F
    Answer = A (page 226)

18. Which of the following units is the oldest?
    A. unit A
    B. unit B
    C. unit C
    D. unit F
    Answer = D (page 226)

19. Which of the following events happened most recently?
    A. deposition of unit C
    B. deposition of unit D
    C. deposition of unit F
    D. tilting of units C, D, E, and F
    Answer = D (page 226)

20. Which of the following statements is true?
    A. The granite is younger than unit B
    B. Unit B is younger than the granite.
    C. Unit B and the granite are the same age.
    D. The relative ages of the granite and unit B can not be determined from the information given.
    Answer = D (page 226)

21. The horizontal line labeled X-X' is a(n) _____.
    A. fault
    B. angular unconformity
    C. contour
    D. cross bed
    Answer = B (page 223)

22. Which of the following statements is false?
    A. Unit A was deposited after unit B was deposited.
    B. Erosion took place prior to deposition of unit B.
    C. Unit C is younger than unit A.
    D. Unit E is older than unit B.
    Answer = C (page 226)

23.   Which of the following occurred most recently?
        A.   deposition of unit A
        B.   deposition of unit D
        C.   emplacement of unit G
        D.   cannot tell from information given
    Answer = D (page 226)

24.   Which of the following is used by geologists to determine the relative ages in a rock sequence?
        A.   stratigraphy
        B.   fossils
        C.   cross-cutting relations
        D.   all of these
    Answer = D (page 225)

25.   When did the Earth's basic structure—the core, mantle, and crust—form?
        A.   during the Archean
        B.   during the Phanerozoic
        C.   during the Proterozoic
        D.   during the Paleozoic
    Answer = A (page 227)

26.   The Jurassic is a geologic _____.
        A.   eon
        B.   epoch
        C.   era
        D.   period
    Answer = D (page 227)

27.   Cenozoic means _____.
        A.   recent life
        B.   age of man
        C.   100 million years
        D.   ice age
    Answer = A (page 227)

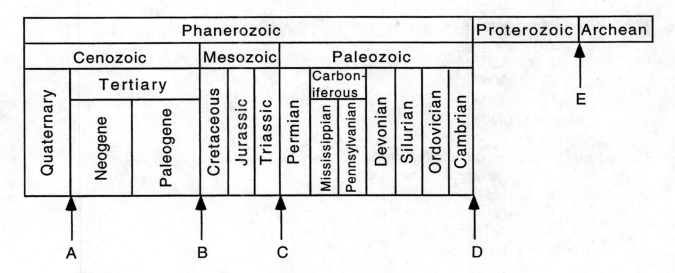

28. The age estimate for point A is _____ million years.
   A. 0.16
   B. 1.6
   C. 16
   D. 160
   Answer = B (page 227)

29. The age estimate for point B is _____ million years.
   A. 35
   B. 65
   C. 245
   D. 570
   Answer = B (page 227)

30. The age estimate for point C is _____ million years.
   A. 35
   B. 65
   C. 245
   D. 570
   Answer = C (page 227)

31. The age estimate for point D is _____ million years.
   A. 65
   B. 245
   C. 570
   D. 2500
   Answer = C (page 227)

32.   The date for point E is _____ million years.
         A.   570
         B.   2500
         C.   3000
         D.   4500
         Answer = C (page 227)

33.   Most periods in the geologic time scale are named for _____.
         A.   geographic localities
         B.   fossils
         C.   catastrophic events
         D.   paleontologists
         Answer = A (page 227)

34.   Who wrote *Principles of Geology*, the first and most influential geology textbook in the 19th century?
         A.   Charles Darwin
         B.   James Hutton
         C.   Alfred Wegener
         D.   Charles Lyell
         Answer = D (page 228)

35.   Which of the following scientists was not associated with important discoveries involving radioactivity?
         A.   Henri Becquerel
         B.   Marie Sklodowska-Curie
         C.   Charles Darwin
         D.   Ernest Rutherford
         Answer = C (page 228)

36.   A rock formed with 1,000 atoms of a radioactive parent element, but only contains 250 radioactive parent atoms today. If the half life for the radioactive element is one million years, how old is the rock?
         A.   250,000 years old
         B.   500,000 years old
         C.   1,000,000 years old
         D.   2,000,000 years old
         Answer = D (page 229)

37.   What scientific avenue of investigation gave scientists the best estimate of the age of the Earth?

      A.   paleontological dating
      B.   archeological dating
      C.   radiometric dating
      D.   carbon dating

Answer = C (page 229)

38.   The stratigraphically highest formation exposed at the Grand Canyon is the _____.

      A.   Bright Angel Shale
      B.   Redwall Limestone
      C.   Kaibab Limestone
      D.   Vishnu Schist

Answer = C (page 231)

39.   Radiometric dating is possible if a rock contains a measurable amount of _____.

      A.   only parent isotopes
      B.   only daughter isotopes
      C.   both parent and daughter isotopes
      D.   either parent or daughter isotopes

Answer = C (page 232)

40.   What instrument is used to count isotopes for radiometric dating?

      A.   mass spectrometer
      B.   electron microprobe
      C.   petrologic microscope
      D.   ion microprobe

Answer = A (page 232)

41.   Radiometric dating is least useful for dating _____ rocks.

      A.   granitic
      B.   basaltic
      C.   sedimentary
      D.   metamorphic

Answer = C (page 232)

42.   Which of the following statements regarding radiometric dating is false?

      A.   Rubidium is incorporated into minerals as they crystallize from a melt.
      B.   Argon-40 is the daughter isotope of potassium-40.
      C.   Carbon-14 is useful for dating materials up to 70,000 years old.
      D.   Sedimentary rocks can be dated more easily than igneous rocks.

Answer = D (page 232)

43.   Which of the following radioactive isotopes has the longest half-life?
    A.   rubidium-87
    B.   potassium-40
    C.   carbon-14
    D.   uranium-238
Answer = A (page 233)

44.   Uranium-238 decays to which of the following daughter products?
    A.   rubidium-87
    B.   uranium-235
    C.   lead-206
    D.   nitrogen-14
Answer = C (page 233)

45.   Which of the following radioactive isotopes is the most useful for dating very young (<10,000 year old) wood and charcoal.
    A.   rubidium-87
    B.   carbon-14
    C.   potassium-40
    D.   uranium-238
Answer = B (page 233)

46.   The radon environmental hazard is associated with the radioactive decay of _____.
    A.   carbon-14
    B.   potassium-40
    C.   rubidium-87
    D.   uranium-238
Answer = D (page 234)

47.   Which of the following areas has rock and/or soil with uranium levels high enough to produce potentially hazardous concentrations of radon?
    A.   the Pacific Coast
    B.   Arizona
    C.   eastern Texas
    D.   the American mid-west
Answer = D (page 235)

48.   If the Earth's history was compressed into a single calendar year, complex organisms (including those with shells) first appeared _____.
    A.   in February
    B.   on June 13
    C.   on October 25
    D.   on December 7
Answer = C (page 237)

49.   Human beings (homo sapiens) evolved during which geologic era?
        A.   Cenozoic
        B.   Mesozoic
        C.   Paleozoic
        D.   Precambrian
    Answer = A (page 238)

50.   Mammals became abundant during the _____ Era.
        A.   Cenozoic
        B.   Mesozoic
        C.   Paleozoic
        D.   Precambrian
    Answer = A (page 238)

10

Folds, Faults, and Other Records of Rock Deformation

1. How many measurements does it take to describe the orientation of a layer of rock exposed at a given location?
    A. 1
    B. 2
    C. 3
    D. 4
    Answer = B (page 245)

2. The dip of a unit represents the _____.
    A. direction of intersection of the rock layer and a horizontal surface
    B. part of the unit which has been eroded
    C. angle at which the bed inclines from the horizontal
    D. tilt of the rock unit before deformation
    Answer = C (page 246)

3. If you are flying in an airplane and you look down at the landscape, you are seeing a _____ view of the earth.
    A. map
    B. cross-sectional
    C. lateral
    D. horizontal
    Answer = A (page 246)

4.   If you look at a vertical face of a cliff or a roadcut, you are seeing a _____ view.
      A.   map
      B.   cross-sectional
      C.   lateral
      D.   horizontal
Answer = B (page 246)

5.   The angle at which a sedimentary bed is inclined from horizontal is called the _____.
      A.   anticline
      B.   strike
      C.   syncline
      D.   dip
Answer = D (page 246)

6.   Which of the following is not a tectonic force responsible for folding or faulting rocks?
      A.   transform force
      B.   compressive force
      C.   tensional force
      D.   shearing force
Answer = A (page 247)

7.   Which of the following types of tectonic forces tends to push two sides of a body in opposite directions so that they slide horizontally past one another?
      A.   tensional forces
      B.   shearing forces
      C.   compressive forces
      D.   all of these
Answer = B (page 247)

8.   What type of forces dominate at convergent plate margins?
      A.   tensional forces
      B.   shearing forces
      C.   compressive forces
      D.   all of these
Answer = C (page 247)

9.   What type of forces dominate at divergent plate margins?
      A.   tensional forces
      B.   shearing forces
      C.   compressive forces
      D.   all of these
Answer = A (page 247)

10. At convergent plate boundaries, one would expect to find _____.
    A. folds
    B. faults
    C. both folds and faults
    D. neither folds nor faults
    Answer = C (page 247)

11. Whether a marble deforms brittlely or ductilely depends upon _____.
    A. temperature
    B. pressure
    C. both temperature and pressure
    D. neither temperature nor pressure
    Answer = C (pages 247–248)

12. Which of the following statements about rock deformation is false?
    A. Deep crustal rocks are more likely to deform ductilely than shallow crustal rocks.
    B. Hotter rocks are more likely to deform ductilely than cooler rocks.
    C. Most sedimentary rocks are more deformable than igneous rocks.
    D. Rocks under low confining pressure are more likely to deform ductilely than rocks under high confining pressure.
    Answer = D (page 248)

13. The biggest difference between rock deformation experiments conducted in a laboratory and rock deformation that occurs naturally is that _____.
    A. the temperatures are much lower in laboratory experiments than in nature
    B. the pressures are much lower in laboratory experiments than in nature
    C. the time of deformation is much shorter in laboratory experiments than in nature
    D. real rocks are not used in laboratory experiments as they are in nature
    Answer = C (page 248)

14. Laboratory experiments indicate that _____.
    A. most igneous rocks are more deformable than most sedimentary rocks
    B. most igneous rocks are less deformable than most sedimentary rocks
    C. basement rocks are more ductile than young sediments
    D. young sediments are very brittle and not easily deformed
    Answer = B (page 248)

15.   A sample of marble has deformed brittlely during a laboratory experiment. If we wish our next sample of marble to deform plastically rather than brittlely, we should conduct the next experiment at _____.
    A.   lower temperatures and lower confining pressures
    B.   lower temperatures and higher confining pressures
    C.   higher temperatures and lower confining pressures
    D.   higher temperatures and higher confining pressures
Answer = D (page 248)

16.   Which of the following conditions would favor folding rather than faulting?
    A.   low temperatures and low confining pressures
    B.   low temperatures and high confining pressures
    C.   high temperatures and low confining pressures
    D.   high temperatures and high confining pressures
Answer = D (page 248)

17.   Which factor does not affect whether folding is severe or gentle?
    A.   the magnitude of the applied forces
    B.   the length of time force was applied
    C.   the age of the rock units
    D.   the ability of the rock units to resist deformation
Answer = C (page 249)

18.   Upfolds, or arches, of layered rock are called _____.
    A.   anticlines
    B.   faults
    C.   synclines
    D.   unconformities
Answer = A (page 249)

19.   The two sides of a fold are called its _____.
    A.   anticlines
    B.   synclines
    C.   limbs
    D.   axial planes
Answer = C (page 249)

20.   A broad circular or oval upward bulge of rock layers is called a(n) _____.
    A.   anticline
    B.   basin
    C.   syncline
    D.   dome
Answer = D (page 250)

21.   If the sedimentary rocks on a geologic map form a zigzag pattern, the underlying structure probably consists of _____.
          A.   horizontal anticlines and synclines
          B.   plunging anticlines and synclines
          C.   domes and basins
          D.   dip-slip and strike-slip faults
       Answer = B (page 250)

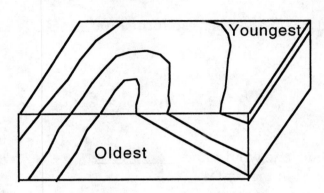

22.   The structure shown in the diagram above is a(n) _____.
          A.   anticline
          B.   syncline
          C.   basin
          D.   dome
       Answer = A (page 251)

23.   The structure shown in the diagram above is _____.
          A.   horizontal and symmetric
          B.   horizontal and asymmetric
          C.   plunging and symmetric
          D.   plunging and asymmetric
       Answer = D (page 251)

24.   An overturned fold is characterized by _____.
          A.   two limbs at right angles to one another
          B.   two limbs dipping in the same direction, with one tilted beyond vertical
          C.   two limbs dipping in opposite directions
          D.   two limbs not parallel to one another
       Answer = B (page 251)

25.   What type of structure is characterized by rock layers which dip radially away from a central point?
       A.   an anticline
       B.   a basin
       C.   a syncline
       D.   a dome
       Answer = D (page 252)

26.   The weight of sediments deposited in a shallow sea can form a(n) _____.
       A.   anticline
       B.   syncline
       C.   basin
       D.   dome
       Answer = C (page 253)

27.   If erosion stripped the top off a dome, one would find _____.
       A.   the oldest rocks exposed in the center
       B.   the youngest rocks exposed in the center
       C.   a linear pattern of rock outcrops
       D.   A and C
       Answer = A (page 253)

28.   Oil can be trapped at the top of a dome if _____.
       A.   there is a nearby basin
       B.   there is an impermeable layer at the top of the dome
       C.   the dome has been eroded
       D.   the dome is part of an adjoining anticline
       Answer = B (page 253)

29.   Where do basins form?
       A.   where parts of the crust have been heated and subsequently cool and contract
       B.   where tensional forces stretch the crust
       C.   where thick piles of sediment are deposited
       D.   all of these
       Answer = D (page 253)

30.   The Black Hills of South Dakota are a good example of a(n) _____.
       A.   anticline
       B.   basin
       C.   syncline
       D.   dome
       Answer = D (page 253)

31.   Which of the following statements about anticlines is true?
      A.   The oldest rocks occur in the center and the limbs dip toward the center.
      B.   The oldest rocks occur in the center and the limbs dip away from the center.
      C.   The youngest rocks occur in the center and the limbs dip toward the center.
      D.   The youngest rocks occur in the center and the limbs dip away from the center.
      Answer = B (page 254)

32.   What types of tectonic forces cause faulting?
      A.   compressive forces
      B.   tensional forces
      C.   shearing forces
      D.   all of these
      Answer = D (page 255)

33.   What type of fault is characterized by horizontal movement?
      A.   a normal fault
      B.   a reverse fault
      C.   a strike-slip fault
      D.   all of these
      Answer = C (page 256)

34.   Which of the following is an example of a fault where the motion is primarily horizontal?
      A.   a strike-slip fault
      B.   a right-lateral fault
      C.   a transform fault
      D.   all of these
      Answer = D (page 256)

35.   What types of faults are associated with shearing forces?
      A.   normal faults
      B.   reverse faults
      C.   strike-slip faults
      D.   all of these
      Answer = C (page 256)

36.   What type of fault is characterized by the rocks above the fault plane moving down-
       ward relative to the rocks below the fault plane?
              A.   a normal fault
              B.   a strike-slip fault
              C.   a reverse fault
              D.   all of these
       Answer = A (page 256)

37.   A fault plane strikes north-south and dips steeply to the west. Geologic observations
       indicate that most of the fault movement was vertical and that Mesozoic rocks occur
       east of the fault and Paleozoic rocks occur west of the fault. What type of fault is
       this?
              A.   a right-lateral fault
              B.   a left-lateral fault
              C.   a normal fault
              D.   a reverse fault
       Answer = D (page 256)

38.   What type of fault is characterized by movement both along slip and along dip?
              A.   strike-slip fault
              B.   oblique-slip fault
              C.   normal fault
              D.   reverse fault
       Answer = B (page 256)

39.   An oblique-slip fault suggests _____ .
              A.   tensional forces only
              B.   compressive forces only
              C.   shear forces only
              D.   shear forces combined with compressive or tensional forces
       Answer = D (page 256)

40.   Strike-slip faults _____ .
              A.   have primarily horizontal displacement
              B.   have primarily vertical displacement
              C.   have no appreciable displacement
              D.   are low-angle reverse faults
       Answer = A (page 256)

41.   Which two measurements describe the orientation of a fault plane at a given location?
      A.   axis and plane
      B.   strike and dip
      C.   lateral and thrust
      D.   trend and plunge
Answer = B (page 256)

42.   Dip-slip faults are associated with _____ forces.
      A.   shearing
      B.   tensional
      C.   compressive
      D.   both tensional and compressive
Answer = D (page 256)

43.   What type of fault is a thrust fault?
      A.   a low-angle normal fault
      B.   a low-angle reverse fault
      C.   a low-angle strike-slip fault
      D.   a low-angle oblique fault
Answer = B (page 257)

44.   Overthrusts are caused by large-scale _____ forces.
      A.   shear
      B.   tensional
      C.   compressive
      D.   shear forces combined with tensional or compressive forces
Answer = C (page 257)

45.   Which of the following geologic features is formed in a region affected by tensional tectonic forces?
      A.   an anticline
      B.   a thrust fault
      C.   a strike-slip fault
      D.   a rift valley
Answer = D (page 258)

46.   The Red Sea is an example of a(n) _____.
      A.   anticline
      B.   strike-slip fault
      C.   rift valley
      D.   thrust fault
Answer = C (page 258)

In the following geologic map, units A, B, and C are sedimentary rocks; unit A is the oldest and unit C is the youngest. The sedimentary rocks are cut by a fault, indicated by the bold line, that dips 60° to the northwest.

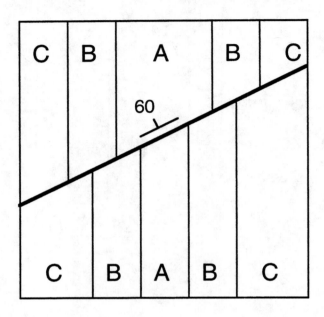

47.   What type of structure is shown in the geologic map?
           A.   a faulted syncline
           B.   a faulted anticline
           C.   a folded strike-slip fault
           D.   the structure cannot be determined from the information given
       Answer = B (page 250)

48.   Which way do the sedimentary layers dip?
           A.   toward the east
           B.   toward the west
           C.   toward the center
           D.   toward the east and west
       Answer = D (page 250)

49.   What type of fault is depicted on the geologic map?
           A.   a normal fault
           B.   a reverse fault
           C.   a right-lateral fault
           D.   a left-lateral fault
       Answer = B (page 256)

50.   Why is unit A wider north of the fault than it is south of the fault?
     A.   Deeper levels of the structure are exposed on the north side of the fault.
     B.   Faulting has thinned unit A south of the fault.
     C.   Unit A had a variable thickness prior to faulting.
     D.   Erosion has removed most of unit A south of the fault.
   Answer = A (page 250)

11

Mass Wasting

1.   Loose, uncemented geologic material is said to be _____.
         A.   liquefied
         B.   crystallized
         C.   unconsolidated
         D.   consolidated
      Answer = C (page 266)

2.   What is the dominant force that causes mass movement?
         A.   tidal force
         B.   seismic energy
         C.   gravity
         D.   wind
      Answer = C (page 266)

3.   Which of the following processes is not strongly influenced by gravity?
         A.   the flow of glacial ice
         B.   the movement of landslides and debris flows
         C.   the movement of water in streams
         D.   all of these are strongly influenced by gravity
      Answer = D (page 266)

4.   Which of the following is the most important factor in causing mass movements?
         A.   temperature
         B.   pressure
         C.   water content
         D.   bulk composition
      Answer = C (page 266)

5. Which of the following does not promote mass movement?
    A.   steep slopes
    B.   forest fires
    C.   heavy rainfall
    D.   all of these promote mass wasting
    Answer = D (page 266)

6. The process by which masses of rock and soil move downhill under the influence of gravity is called _____.
    A.   landsliding
    B.   mass wasting
    C.   hydraulic action
    D.   solifluction
    Answer = B (page 266)

7. Approximately how many people were killed in the mudflow of volcanic ash that swept down the flank of Nevada del Ruiz in Columbia in 1985?
    A.   20,000
    B.   2,000
    C.   200
    D.   20
    Answer = A (page 266)

8. A hill consisting of loose, dry sand that slopes at the angle of repose and has no vegetation _____.
    A.   is stable unless oversteepened by excavation
    B.   may flow if it becomes saturated with water
    C.   will be more stable if vegetation takes root on the hill
    D.   all of these
    Answer = D (pages 266–268)

9. Which of the following statements is false?
    A.   Round debris forms steeper slopes than angular debris.
    B.   Large debris forms steeper slopes than small debris.
    C.   Dry debris forms steeper slopes than water-saturated debris.
    D.   Moist debris forms steeper slopes than dry debris.
    Answer = A (page 267)

10. Which of the following situations is least likely to result in mass movement?
    A.   a steep slope
    B.   a slope with loose material saturated with water
    C.   a slope with abundant vegetation
    D.   a slope consisting of fractured and deformed rock
    Answer = C (page 267)

11.   Which of the following has the steepest angle of repose?
         A.   fine quartz sand
         B.   coarse quartz sand
         C.   angular quartz pebbles
         D.   all of these have the same angle of repose
       Answer = C (page 267)

12.   The characteristic slope of a pile of dry sand is called the _____.
         A.   angle of repose
         B.   strike
         C.   consolidation factor
         D.   dip
       Answer = A (page 267)

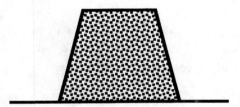

13.   This illustration most likely depicts a cross section through a pile of _____.
         A.   dry sand
         B.   moist sand
         C.   water-saturated sand
         D.   water-oversaturated sand
       Answer = B (page 267)

14.   Which of the following factors affects the maximum angle at which a slope of loose material is stable?
         A.   the amount of moisture
         B.   the shape of the particles
         C.   the size of the particles
         D.   all of these
       Answer = D (page 267)

15.   Which of the following situations is most likely to undergo mass wasting?
         A.   a dry, moderate slope of unconsolidated material
         B.   a wet, moderate slope of unconsolidated material
         C.   a dry, steep slope of unconsolidated material
         D.   a wet, steep slope of unconsolidated material
       Answer = D (page 267)

16. Damp sand has a higher angle of repose than dry sand because of _____.
     A.   cementation
     B.   surface tension
     C.   partial melting
     D.   dissolution
     Answer = B (page 268)

17. As the amount of water in a pile of quartz sand increases, the angle of repose will
     _____.
     A.   increase
     B.   decrease
     C.   first increase and then decrease
     D.   not change
     Answer = C (page 268)

18. Surface tension is greatest when _____.
     A.   sand is dry
     B.   sand is moist, but not saturated with water
     C.   sand is saturated with water
     D.   sand is oversaturated with water
     Answer = B (page 268)

19. Which of the following rock types form the steepest slopes?
     A.   granite
     B.   shale
     C.   volcanic ash beds
     D.   all of these rock types form steep slopes
     Answer = A (page 269)

20. Which of the following slopes is least stable?
     A.   a slope where the sedimentary layers dip parallel to the slope
     B.   a slope where the sedimentary layers are horizontal
     C.   a slope where the sedimentary layers dip perpendicular to the slope
     D.   all of these slopes have the same stability
     Answer = A (page 270)

21. Which of the following can trigger a landslide?
     A.   an earthquake
     B.   a heavy rainstorm
     C.   removal of material from the base of a slope
     D.   all of these
     Answer = D (page 270)

22.   Most of the damage associated with the 1964 Anchorage, Alaska earthquake was caused by _____.
        A.   ground shaking during the earthquake
        B.   a volcanic eruption triggered by the earthquake
        C.   a tsunami (tidal wave) triggered by the earthquake
        D.   landslides triggered by the earthquake
Answer = D (page 270)

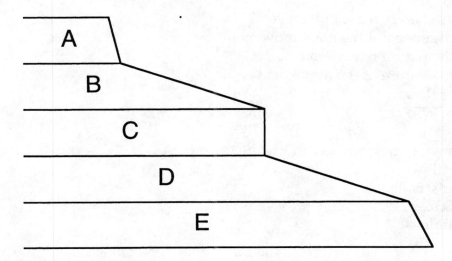

23.   In the cross section through the upper part of the Grand Canyon depicted here, units B and D are most likely composed of _____.
        A.   granite
        B.   limestone
        C.   sandstone
        D.   shale
Answer = D (page 270)

24.   Which of the following processes was the major reason so many landslides occurred during the 1964 Anchorage, Alaska earthquake?
        A.   motion along the fault oversteepened slopes
        B.   water-saturated sandy layers became liquefied by the ground shaking
        C.   the earthquake tilted the rock layers downhill
        D.   the earthquake caused water to accumulate in the soil
Answer = B (page 272)

25.   During an earthquake, water-saturated sand can behave like a liquid, a process called
       _____.
              A.   slurrification
              B.   unconsolidation
              C.   liquefaction
              D.   solifluction
       Answer = C (page 272)

26.   How do geologists classify mass movements?
              A.   by the speed of the mass movement
              B.   by the nature of material
              C.   by the nature of the movement
              D.   all of these
       Answer = D (page 273)

27.   Which of the following mass movements is the fastest?
              A.   mudflow
              B.   debris avalanche
              C.   soil creep
              D.   earthflow
       Answer = B (page 273)

Dominant Material	Nature of Motion	Slow (< 1 cm/yr)	Moderate (1–5 km/hr)	Fast (>5 km/hr)
Rock	Flow	a	b	c
Rock	Slide or fall	d	e	f
Unconsolidated	Flow	g	h	i
Unconsolidated	Slide or fall	j	k	l

28.   How would a debris avalanche be classified in this table?
              A.   box c
              B.   box f
              C.   box i
              D.   box l
       Answer = C (page 273)

29.   How would a mudflow be classified in this table?
       A.   box a
       B.   box b
       C.   box g
       D.   box h
Answer = D (page 273)

30.   How would a slump be classified in this table?
       A.   box b
       B.   box e
       C.   box h
       D.   box k
Answer = D (page 273)

31.   How would creep be classified in this table?
       A.   box a
       B.   box e
       C.   box g
       D.   box k
Answer = C (page 273)

32.   How would a rock avalanche be classified in this table?
       A.   box b
       B.   box c
       C.   box e
       D.   box f
Answer = B (page 273)

33.   The accumulation of rocks at the base of a cliff is called _____.
       A.   a dune
       B.   soil creep
       C.   an alluvial fan
       D.   talus
Answer = D (page 274)

34.   Which of the following statements about mudflows is false?
       A.   Mudflows tend to move slower than debris flows.
       B.   Mudflows are most common in semi-arid regions.
       C.   Mudflows contain large amounts of water.
       D.   Mudflows can carry large boulders.
Answer = A (page 276)

35.   What is the difference between an earthflow and a debris flow?
       A.   Earthflows travel faster than debris flows.
       B.   Earthflows travel slower than debris flows.
       C.   Earthflows consist of finer material than debris flows.
       D.   Earthflows consist of coarser material than debris flows.
       Answer = C (page 276)

36.   How fast is soil creep ?
       A.   1 to 10 millimeters per year
       B.   1 to 10 meters per year
       C.   1 kilometer per hour
       D.   over 10 kilometers per hour
       Answer = A (page 276)

37.   Which of the following types of mass movements could a person not outrun ?
       A.   slump
       B.   debris avalanche
       C.   soil creep
       D.   debris flow
       Answer = B (page 276)

38.   Why are mudflows and debris avalanches common on volcanic slopes?
       A.   because there is abundant unconsolidated volcanic ash
       B.   earthquakes associated with volcanic eruptions trigger mass move-
               ments
       C.   volcanic eruptions trigger melting of ice and snow
       D.   all of these
       Answer = D (page 277)

39.   Which of the following mass wasting events is best classified as a flow?
       A.   the 1980 Mt. St. Helens debris avalanche in Washington
       B.   the 1983 Spanish Fork Canyon landslide in Utah
       C.   the 1925 Gros Ventre landslide near Jackson Hole, Wyoming
       D.   the 1963 Vaiont landslide in the Italian Alps
       Answer = A (page 277)

40.   Which of the following mass movement events was triggered by an earthquake?
       A.   the Spanish Fork Canyon landslide, Utah
       B.   the Gros Ventre landslide, Wyoming
       C.   the Nevado Huascaran ice-debris avalanche, Peruvian Andes
       D.   all of these were triggered by earthquakes
       Answer = C (page 277)

41.   Approximately how many people die in the U.S. each year as a result of landslides?
      A.   usually none
      B.   1 to 10
      C.   25 to 50
      D.   more than 100
Answer = C (page 278)

42.   A slow slide of unconsolidated material that travels as a unit is called _____.
      A.   soil creep
      B.   a debris flow
      C.   a slump
      D.   a rockslide
Answer = C (page 279)

43.   Solifluction may occur when _____.
      A.   the surface soil layer freezes while the deeper soil remains unfrozen
      B.   the surface soil layer thaws while the deeper soil remains frozen
      C.   the surface soil layer and the deeper soil both freeze
      D.   the surface soil layer and the deeper soil both thaw
Answer = B (page 279)

44.   A slump is _____.
      A.   a rock flow
      B.   a rock slide
      C.   a flow of unconsolidated material
      D.   a slide of unconsolidated material
Answer = D (page 279)

45.   Which of the following statements best describes a slump?
      A.   slow downhill movement of unconsolidated material moving as a unit
      B.   slow downhill movement of unconsolidated material moving like a fluid
      C.   rapid downhill movement of unconsolidated material moving as a unit
      D.   rapid downhill movement of unconsolidated material moving like a fluid
Answer = A (page 279)

46.   _____ only occurs in permafrost regions.
      A.   Landsliding
      B.   Soil creep
      C.   Liquefaction
      D.   Solifluction
Answer = D (page 279)

47.   Why are few mass movements preserved in the geologic record?
      A.   Mass movement deposits erode easily.
      B.   Mass movements have only occurred recently due to human activities.
      C.   Mass movements were rare in the past due to low rainfall.
      D.   Mass movement deposits are destroyed by diagenesis.
    Answer = A (page 280)

48.   What of the following was not a contributing factor in the 1925 Gros Ventre slide near Jackson Hole, Wyoming?
      A.   the orientation of the sedimentary layers with respect to the slope
      B.   undercutting of the slope by construction activities
      C.   heavy rains and melting snow
      D.   all of these were contributing factors
    Answer = B (page 281)

49.   What was the maximum velocity of the 1925 Gros Ventre landslide in Wyoming?
      A.   about one mile per hour
      B.   about ten miles per hour
      C.   about fifty miles per hour
      D.   about two hundred miles per hour
    Answer = C (page 281)

50.   Mass movements frequently occur at which of the following plate tectonic settings?
      A.   convergent plate boundaries
      B.   divergent plate boundaries
      C.   transform plate boundaries
      D.   all of the above
    Answer = D (page 282)

12

The Hydrologic Cycle
and Groundwater

1.   Which of the following reservoirs contains the most water?
            A.   atmosphere
            B.   biosphere
            C.   groundwater
            D.   lakes and rivers
        Answer = C (page 288)

2.   How much of the Earth's water is stored in underground aquifers?
            A.   less than 1%
            B.   about 5%
            C.   about 20%
            D.   about 50%
        Answer = A (page 288)

3.   There is approximately _____ cubic kilometers of water on Earth.
            A.   one million
            B.   one billion
            C.   one trillion
            D.   one quadrillion
        Answer = B (page 288)

4.   What is the process by which water enters the small pore spaces between particles of soil or rock?

    A.   transpiration
    B.   infiltration
    C.   precipitation
    D.   sublimation

Answer = B (page 289)

5.   With respect to the Earth's land surface, which of the following mathematical balances is correct?

    A.   precipitation = evaporation – runoff
    B.   precipitation = runoff – evaporation
    C.   precipitation = evaporation + runoff
    D.   precipitation = evaporation × runoff

Answer = C (page 290)

6.   Which of the following terms is a measure of the amount of water vapor in the air as a proportion of the maximum amount the air could hold at the same temperature?

    A.   dew point
    B.   sublimation point
    C.   evaporation rate
    D.   relative humidity

Answer = D (page 291)

7.   Which of the following statements is true?

    A.   Cool air can hold more water vapor than warm air.
    B.   Warm air can hold more water vapor than cool air.
    C.   Cool air and warm air can hold the same amount of water vapor.
    D.   Air cannot contain water vapor.

Answer = B (page 291)

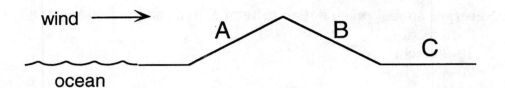

8.   This diagram represents a cross section through a coastal mountain range. Which of the following statements is correct?
     A.   Region A would receive the most precipitation.
     B.   Region B would receive the most precipitation.
     C.   Region C would receive the most precipitation.
     D.   Regions A, B, and C would receive about the same amount of precipitation.
     Answer = A (page 291)

9.   Approximately how much water does a human need each day to survive?
     A.   about 1/4 of a gallon (one quart)
     B.   about 1/2 of a gallon (two quarts)
     C.   about 1 gallon (four quarts)
     D.   about 2 gallons (eight quarts)
     Answer = B (page 292)

10. Per capita domestic water use in the United States is _____ that of western Europe.
     A.   approximately 10 times less than
     B.   approximately 3 times less than
     C.   approximately the same as
     D.   approximately 3 times greater than
     Answer = D (page 292)

11. California imports most of its water supply. What is most (> 80%) of the water used for?
     A.   agriculture
     B.   domestic irrigation (lawns)
     C.   industry
     D.   municipal drinking water
     Answer = A (page 292)

12. Which of the following regions has the highest average annual precipitation?
     A.   northeast United States
     B.   southeast United States
     C.   southwest United States
     D.   central United States
     Answer = B (page 294)

13.   What is the average annual precipitation in Tempe, Arizona [your city, state]?
    A.   0–10 inches
    B.   10–20 inches
    C.   20–30 inches
    D.   30–40 inches
    Answer = A for Tempe, Arizona (page 294)

14.   What is the average annual runoff in central Arizona [your local region]?
    A.   0–1 inch
    B.   1–5 inches
    C.   5–20 inches
    D.   > 20 inches
    Answer = A for Tempe, Arizona (page 294)

15.   Which river carries the most water?
    A.   the Amazon River in South America
    B.   the Ganges River in Asia
    C.   the Congo River in Africa
    D.   the Mississippi River in North America
    Answer = A (page 295)

16.   In the United States, what percentage of the original wetlands have been destroyed?
    A.   less than 5%
    B.   about 10%
    C.   about 25%
    D.   more than 50%
    Answer = D (page 296)

17.   Groundwater represents how much of the world's fresh water?
    A.   about 1%
    B.   about 5%
    C.   about 20%
    D.   about 50%
    Answer = C (page 296)

18.   The percentage of a rock's total volume taken up by pore space is called the _____.
    A.   permeability
    B.   recharge
    C.   aquifer
    D.   porosity
    Answer = D (page 297)

19.   Permeability is _____.
      A.   the ability of a solid to allow fluids to pass through
      B.   the process by which plants release water vapor to the atmosphere
      C.   the amount of water vapor in the air relative to the maximum amount of water vapor the air can hold
      D.   the percentage of pore space in the rock
Answer = A (page 297)

20.   Which of the following rocks has the highest permeability?
      A.   a unfractured shale
      B.   a cemented sandstone
      C.   an uncemented sandstone
      D.   all of these rocks have approximately the same permeability
Answer = C (page 296)

21.   Which of the following materials has the lowest porosity?
      A.   shale
      B.   gravel
      C.   granite
      D.   sandstone
Answer = C (pages 297–298)

22.   Which of the following combinations make for the best groundwater reservoir?
      A.   low permeability and low porosity
      B.   low permeability and high porosity
      C.   high permeability and low porosity
      D.   high permeability and high porosity
Answer = D (page 298)

23.   Which of the following statements about the water table is false?
      A.   The water table changes when discharge is not balanced by recharge.
      B.   The water table is generally flat.
      C.   The water table is above the land surface in lakes.
      D.   The water table is depressed near high volume pumping wells.
Answer = B (page 298)

24.   What is the difference between the saturated and unsaturated zones of groundwater?
      A.   The saturated zone has a higher porosity than the unsaturated zone.
      B.   The saturated zone has a lower porosity than the unsaturated zone.
      C.   The pore spaces in the saturated zone are completely full of water; the pore spaces in the unsaturated zone are not completely full of water.
      D.   The pore spaces in the saturated zone are not completely full of water; the pore spaces in the unsaturated zone are completely full of water.
Answer = C (page 298)

25.   The boundary between the saturated zone and the unsaturated zone is called the
_____.
        A.   water table
        B.   aquifer
        C.   aquiclude
        D.   porosity
Answer = A (page 298)

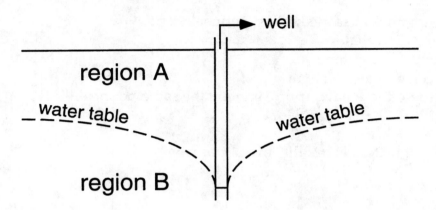

26.   In this diagram region A is the _____.
        A.   discharge zone
        B.   recharge zone
        C.   saturated zone
        D.   unsaturated zone
Answer = D (page 298)

27.   In this diagram region B is the _____.
        A.   discharge zone
        B.   recharge zone
        C.   saturated zone
        D.   unsaturated zone
Answer = C (page 298)

28.   The lowering of the water table near the well is called a(n) _____. Refer to this
diagram to help you.
        A.   aquiclude
        B.   cone of depression
        C.   influent zone
        D.   sinkhole
Answer = B (page 303)

29.   The infiltration of water into the subsurface is called _____.
    A.   influent
    B.   effluent
    C.   discharge
    D.   recharge
Answer = D (page 299)

30.   Influent streams are _____.
    A.   more common in arid regions
    B.   more common in humid regions
    C.   equally common in arid and humid regions
    D.   only found in areas of permafrost
Answer = A (page 299)

WEST   **EAST**

surface

stream

water table

31.   Given the geometry of the surface and water table depicted in this cross section, which of the following statements is true?
    A.   The stream is gaining water from the groundwater.
    B.   The stream is losing water to the groundwater.
    C.   The stream is gaining water from the groundwater on the west side and losing water to the groundwater on the east side.
    D.   The stream and the groundwater are not connected.
Answer = A (page 299)

32.   What is the term for a relatively impermeable geologic unit?
    A.   an artesian
    B.   an aquiclude
    C.   an aquifer
    D.   none of these
Answer = B (page 300)

33.   A confined aquifer is recharged by _____.
    A.   precipitation where the aquifer unit crops out
    B.   water draining down from overlying streams
    C.   precipitation draining down through the unsaturated zone
    D.   all of these
Answer = A (page 300)

34.   Excessive pumping in relation to recharge can cause _____.
    A.   the water table to decline
    B.   a cone of depression to form
    C.   the well to go dry
    D.   all of these
Answer = D (page 301)

35.   Which of the following is not caused by overpumping groundwater?
    A.   intrusion of salt water into coastal aquifers
    B.   depletion of an aquifer
    C.   raising of the land surface
    D.   development of cracks and fissures at the surface
Answer = C (page 302)

36.   Which of the following statements about groundwater is false?
    A.   The steeper the water-table slope, the faster the groundwater will flow.
    B.   Groundwater moves from areas where the water table is high to areas where the water table is low.
    C.   The higher the permeability of an aquifer, the slower the groundwater will flow.
    D.   Gravity drives the flow of groundwater.
Answer = C (page 303)

37.   How fast does most groundwater move in aquifers?
    A.   a few centimeters per day
    B.   a few meters per day
    C.   a few kilometers per day
    D.   a few kilometers per hour
Answer = A (page 304)

38.   Most groundwater withdrawn in the United States is used for _____.
    A.   industry
    B.   irrigation
    C.   drinking water
    D.   swimming pools
Answer = B (page 305)

39.   In what type of rock do most caves form?
    A.   granite
    B.   shale
    C.   limestone
    D.   sandstone
Answer = C (page 305)

40.   In which of the following regions is groundwater being significantly depleted?
        A.   Arizona
        B.   western Texas
        C.   California
        D.   all of the above
        Answer = D (page 306)

41.   The Ogallala aquifer in the southern plains consists of _____?
        A.   basalt
        B.   limestone
        C.   sand and gravel
        D.   shale
        Answer = C (page 306)

42.   Which of the following statements about karst topography is false?
        A.   Karst topography contains many sinkholes.
        B.   Karst topography forms from freezing and thawing of groundwater.
        C.   Karst topography does not have a normal surface drainage system consisting of small and large rivers.
        D.   Karst topography forms in regions where subsurface limestone is dissolved by groundwater.
        Answer = B (page 307)

43.   Stalactites and stalagmites in caves are composed of _____.
        A.   quartz
        B.   feldspar
        C.   calcite
        D.   halite
        Answer = C (page 307)

44.   Sinkholes are a possible danger in regions underlain by what type of bedrock?
        A.   granite
        B.   sandstone
        C.   shale
        D.   limestone
        Answer = D (page 307)

45.   "Hard" water contains large amounts of _____.
        A.   dissolved lead
        B.   dissolved sodium
        C.   dissolved calcium carbonate
        D.   dissolved silica
        Answer = C (page 308)

46.   Water that is good enough to drink is called _____.
         A.   potable water
         B.   groundwater
         C.   surface water
         D.   artesian water
       Answer = A (page 309)

47.   Which of the following aquifers would be best for removing bacteria from ground-
       water?
         A.   limestone
         B.   sand
         C.   coarse gravel
         D.   none of these can remove effectively bacteria from groundwater
       Answer = B (page 309)

48.   Which of the following sources can contaminate an aquifer?
         A.   landfills
         B.   agricultural regions
         C.   gas stations
         D.   all of these
       Answer = D (page 310)

49.   As one goes deeper in the Earth's crust, _____.
         A.   the porosity increases and the concentration of dissolved minerals
               increases
         B.   the porosity increases and the concentration of dissolved minerals
               decreases
         C.   the porosity decreases and the concentration of dissolved minerals
               increases
         D.   the porosity decreases and the concentration of dissolved minerals
               decreases
       Answer = C (page 311)

50.   Most of the water coming out of continental hot springs is _____.
         A.   meteoric water
         B.   magmatic water
         C.   seawater
         D.   metamorphic water
       Answer = A (page 312)

13

Rivers: Transport to the Oceans

1. Which of the following processes is the most important erosional force?
    A.   streams
    B.   glaciers
    C.   wind
    D.   waves
    Answer = A (page 318)

2. How much dissolved materials do rivers transport each year?
    A.   approximately 2–4 thousand tons
    B.   approximately 2–4 million tons
    C.   approximately 2–4 billion tons
    D.   approximately 2–4 trillion tons
    Answer = C (page 318)

3. Since humans began to actively affect their environment, how much has sediment transport by stream systems increased?
    A.   10%
    B.   50%
    C.   100%
    D.   200%
    Answer = C (page 318)

4. Which of the following flows is most likely to be turbulent?
    A.   slow flow in a shallow channel
    B.   fast flow in a shallow channel
    C.   slow flow in a deep channel
    D.   fast flow in a deep channel
    Answer = D (page 319)

5.   Which of the following statements about fluid flow is false?
    A.   As the velocity of a stream increases, laminar flow may change to turbulent flow.
    B.   The viscosity of most fluids increases as temperature increases.
    C.   Most streams and rivers are turbulent.
    D.   The more viscous the fluid, the more likely the flow is laminar.
Answer = B (page 319)

6.   What type of flow can transport gravel and cobbles?
    A.   laminar flow
    B.   turbulent flow
    C.   both A and B
    D.   neither A nor B
Answer = B (page 319)

7.   What type of material is most likely to be transported as suspended load?
    A.   clay particles
    B.   sand
    C.   gravel
    D.   All of these are equally likely to be transported as suspended load.
Answer = A (page 319)

8.   Which of the following statements about streams is false?
    A.   For the same discharge, laminar flows generally carry more sediment than turbulent flows.
    B.   Faster currents can carry larger particles than slower currents.
    C.   Smaller particles settle more slowly than larger particles.
    D.   The base level is the lowest level to which a stream can erode.
Answer = A (page 319)

9.   Particles that roll and slide along the river bottom are called _____.
    A.   bed load
    B.   suspended load
    C.   either bed load or suspended load depending on the particle size
    D.   neither bed load nor suspended load
Answer = A (page 319)

10.   Suspended load includes material _____.
    A.   rolling along the bottom of the stream
    B.   temporarily or permanently suspended in the flow
    C.   deposited on the bottom of the stream
    D.   rolling along the bottom and suspended in the flow
Answer = B (page 319)

11. Gravel-sized particles are transported by rivers as _____.
    A.   suspended load
    B.   bed load
    C.   dissolved load
    D.   all of these
Answer = B (page 319)

12. The speed at which suspended particles become part of the bed of a stream is called the _____.
    A.   discharge velocity
    B.   capacity
    C.   settling velocity
    D.   bed rate
Answer = C (page 319)

13. The intermittent jumping motion of sand grains along a river bottom is called _____.
    A.   saltation
    B.   rippling
    C.   suspension
    D.   meandering
Answer = A (page 320)

14. During turbulent flow, smaller grains will not _____.
    A.   be picked up more frequently than large grains
    B.   jump higher than large grains
    C.   settle more quickly than large grains
    D.   travel further than large grains
Answer = C (page 320)

15. Ripples are commonly _____ high.
    A.   several millimeters
    B.   several centimeters
    C.   several decimeters
    D.   several meters
Answer = B (page 321)

16. Which of the following stream velocities will lead to the largest sand dunes?
    A.   low velocity
    B.   moderate velocity
    C.   high velocity
    D.   very high velocity
Answer = C (page 321)

17. Running water erodes solid rock by _____.
    A.   abrasion
    B.   chemical and physical weathering
    C.   the undercutting action of currents
    D.   all of the above
Answer = D (page 322)

18. Potholes in river bottom bedrock are formed by _____.
    A.   the impact of a large rock moved by a strong current which makes a "crater"
    B.   the grinding action of a pebble or cobble in a swirling eddy
    C.   cascading water from a waterfall which wears away the rock
    D.   none of the above
Answer = B (page 322)

19. Of the choices shown below, the most common cross-sectional river valley profile is _____.

A.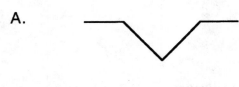

B.

C.

D.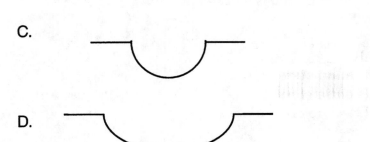

Answer = A (page 323)

20. What is the term for a curved sandbar that forms on the inside curve of a stream?
    A.   meander
    B.   point bar
    C.   dune
    D.   oxbow
Answer = B (page 324)

21.   At a bend in a river, _____ occurs on the outside of the bend and _____ occurs on the inside of the bend.
          A.   erosion . . . deposition
          B.   deposition . . . erosion
          C.   erosion . . . erosion
          D.   deposition. . . deposition
Answer = A (page 324)

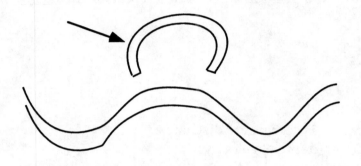

22.   In this figure the arrow is pointing to a(n) _____.
          A.   point bar
          B.   oxbow lake
          C.   sand dune
          D.   natural levee
Answer = B (page 325)

23.   Natural levees are made up of _____.
          A.   silt and clay deposited during a flood
          B.   sand and gravel deposited during a flood
          C.   overlapping point bars
          D.   isolated point bars
Answer = B (page 326)

24.   Which of the following is does not determine whether a stream is straight, braided, or meandering?
          A.   flow volume
          B.   sediment load
          C.   river bank erodibility
          D.   length of river
Answer = D (page 326)

25.   Natural levees are built up by _____.
           A.   beavers
           B.   humans
           C.   floods
           D.   erosion
        Answer = C (page 326)

26.   The volume of water flowing past a point in a given time is called the _____.
           A.   competence
           B.   viscosity
           C.   discharge
           D.   capacity
        Answer = C (page 328)

27.   Which of the following equations is correct for discharge?
           A.   discharge = width/(depth × velocity)
           B.   discharge = (width × depth)/velocity
           C.   discharge = width × depth × velocity
           D.   discharge = (width × velocity)/depth
        Answer = C (page 328)

28.   Which of the following will increase the discharge of a stream?
           A.   increasing the velocity, keeping width and depth constant
           B.   increasing the width, keeping velocity and depth constant
           C.   increasing the depth, keeping velocity and width constant
           D.   all of these
        Answer = D (page 328)

29.   What is the probability of a 50-year flood occurring next year?
           A.   2%
           B.   10%
           C.   50%
           D.   cannot be determined from the information given
        Answer = A (page 329)

30.   For a given river, which of the following floods would have the largest discharge?
           A.   a 5-year flood
           B.   a 20-year flood
           C.   a 100-year flood
           D.   One cannot tell from the information provided.
        Answer = C (page 329)

31. Which of the following statements regarding floods is true?
     A. A 50-year flood is generally of greater magnitude than a 100-year flood.
     B. A 100-year flood has a 10% chance of occurring in any one year.
     C. If there is a 20% probability of a flood of a certain height occurring in any one year, it is called a 5-year flood.
     D. The recurrence interval of a flood of a certain height does not depend upon the width of the floodplain.
     Answer = C (page 329)

32. At any point in a river, the equilibrium between erosion of the streambed and sedimentation in the channel is not controlled by _____.
     A. topography
     B. climate
     C. stream flow
     D. river length
     Answer = D (page 330)

33. The shape of the longitudinal profile of all streams is _____.
     A. a horizontal line
     B. a straight line sloping downstream
     C. a concave upward curve
     D. a concave downward curve
     Answer = C (page 331)

34. If sea level were to rise, the slope of the longitudinal profile of many rivers would _____.
     A. increase
     B. decrease
     C. first increase, then decrease
     D. first decrease, then increase
     Answer = B (page 332)

35. Why is the primary reason sediment is deposited in large cone-shaped deposits at mountain fronts?
     A. Because stream valleys widen abruptly at a mountain front.
     B. Because stream valleys narrow abruptly at a mountain front.
     C. Because stream valleys get much steeper at a mountain front.
     D. Because stream valleys get less steep at a mountain front.
     Answer = A (page 332)

36.   If a dam is built, sediment will _____ on the upstream side of the dam and sediment will _____ on the downstream side of the dam.
        A.   accumulate ... accumulate
        B.   accumulate ... erode
        C.   erode ... accumulate
        D.   erode ... erode
Answer = B (page 333)

37.   River terraces are composed of _____ and form as a result of rapid _____.
        A.   bedrock ... subsidence
        B.   bedrock ... uplift
        C.   floodplain deposits ... subsidence
        D.   floodplain deposits ... uplift
Answer = D (page 334)

38.   Terraces are remnants of former _____.
        A.   floodplains
        B.   rivers
        C.   archeological ruins
        D.   alluvial fans
Answer = A (page 334)

39.   If a stream breaks through a divide and captures drainage from the competing stream, it is called _____.
        A.   competitive capture
        B.   competitive erosion
        C.   stream piracy
        D.   stream capture
Answer = C (page 335)

40.   What is the most common drainage pattern?
        A.   radial drainage
        B.   internal drainage
        C.   trellis drainage
        D.   dendritic drainage
Answer = D (page 336)

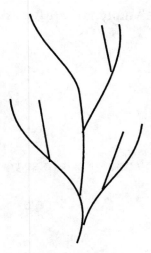

41.   This pattern most closely resembles _____ drainage.
      A.   dendritic
      B.   rectangular
      C.   radial
      D.   trellis
Answer = A (page 336)

42.   Radial drainage forms in regions with _____.
      A.   parallel ridges of resistant rock
      B.   mountain chains
      C.   bedrock with joint fractures
      D.   a central uplift
Answer = D (page 336)

43.   What type of drainage network would you expect to find on a volcano?
      A.   dendritic drainage
      B.   rectangular drainage
      C.   radial drainage
      D.   none of these
Answer = C (page 336)

44.   A _____ drainage pattern develops where rapid weathering along joints in bedrock controls the course of streams
      A.   dendritic
      B.   rectangular
      C.   radial
      D.   concentric
Answer = B (page 336)

45.   What type of drainage would you expect to observe in the Valley and Ridge province
       of the southern Appalachians?
            A.   radial drainage
            B.   internal drainage
            C.   trellis drainage
            D.   dendritic drainage
       Answer = C (page 336)

46.   How far can large rivers, such as the Amazon, maintain a current out to sea?
            A.   meters
            B.   tens of meters
            B.   hundreds of meters
            D.   many kilometers
       Answer = D (page 338)

47.   A delta is made up of sediments _____.
            A.   deposited at the mouth of a river
            B.   deposited on the inside of a meander loop
            C.   deposited at a mountain front
            D.   deposited on the outside of a meander loop
       Answer = A (page 338)

48.   Which of the following beds in a delta consist of thin horizontal layers of mud?
            A.   topset beds
            B.   foreset beds
            C.   bottomset beds
            D.   all of these
       Answer = C (page 338)

49.   Why is the Mississippi delta so large?
            A.   Because the Mississippi River transports a huge amount of sediment.
            B.   Because tides in the Gulf of Mexico are not very strong.
            C.   Because waves in the Gulf of Mexico are not very strong.
            D.   All of these.
       Answer = D (page 340)

50.   The east coast of North America lacks deltas because _____.
            A.   waves and tides are too strong
            B.   no rivers empty out along the east coast of North America
            C.   the Appalachian Mountains are too erosion resistant
            D.   rivers of the east coast have currents that are too weak to carry much
                  sediment
       Answer = A (page 340)

14

Winds and Deserts

1.   Geologic processes powered by the wind are called _____ processes.
        A.   barchan
        B.   coriolis
        C.   eolian
        D.   yardang
     Answer = C (page 346)

2.   Turbulence of air increases in proportion to the _____ of the air flow.
        A.   volume
        B.   velocity
        C.   acceleration
        D.   distance
     Answer = B (page 346)

3.   Wind speeds of 117 kilometers per hour or more constitute a _____.
        A.   moderate to strong breeze
        B.   moderate to strong gale
        C.   whole gale to storm
        D.   hurricane
     Answer = D (page 346)

4.   In the temperate zones of the Earth between 30° and 60° latitude, the prevailing winds come from the _____.
        A.   north
        B.   south
        C.   east
        D.   west
     Answer = D (page 346)

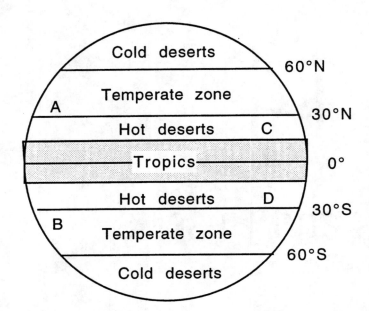

5.   In this figure, the letters A through D represent wind names. Which of the following
      statements regarding the above figure is false?
         A.   A and B represent "westerlies."
         B.   A and C represent "trade winds."
         C.   C and D represent "trade winds."
         D.   None of the choices are false.
      Answer = B (page 347)

6.   Due to the Coriolis effect, any current of air or water is deflected to the _____ in the
      northern hemisphere and to the _____ in the southern hemisphere.
         A.   east ... east
         B.   east ... west
         C.   west ... east
         D.   west ... west
      Answer = B (page 347)

7.   Sand grains saltating in air can jump higher than sand grains saltating in water
      because _____.
         A.   water is less dense than air
         B.   air is less viscous than water
         C.   wind velocity is faster than water velocity
         D.   grains are suspended in air longer than they are in water
      Answer = B (page 348)

8.   Ripples are generally oriented _____ to the current direction.
        A.   parallel
        B.   30 degrees
        C.   45 degrees
        D.   perpendicular
Answer = D (page 348)

9.   In a large dust storm, approximately how much dust can 1 cubic kilometer of air carry?
        A.   1 ton
        B.   10 tons
        C.   100 tons
        D.   1000 tons
Answer = D (page 349)

10.   Fine-grained dust particles from the eruption of Mt. Pinatubo remained in the atmosphere for how long?
        A.   1–2 months
        B.   3–4 months
        C.   1–2 years
        D.   3–4 years
Answer = D (page 349)

11.   White Sands National Monument contains sand dunes made up of _____.
        A.   quartz
        B.   gypsum
        C.   halite
        D.   feldspar
Answer = B (page 351)

12.   Which of the following is an important source of dust?
        A.   volcanic dust from eruptions
        B.   clay minerals from soils
        C.   organic sources, including charcoal, pollen, and bacteria
        D.   all of these
Answer = D (page 351)

13.   What is the primary cause of frosting of the surface of quartz sand grains?
        A.   impacts with other sand grains
        B.   abrasion by air currents
        C.   slow dissolution by dew
        D.   none of these
Answer = C (page 351)

14.   Which of the following statements is true?
     A.   Wind is a more effective erosion agent in wet climates.
     B.   Wind is a more effective erosion agent in dry climates.
     C.   Wind is an equally effective erosion agent in wet and dry climates.
     D.   Wind does not cause erosion.
Answer = B (page 351)

15.   The process by which the ground surface is lowered by wind erosion is called _____.
     A.   deflation
     B.   inflation
     C.   ablation
     D.   none of these
Answer = A (page 351)

16.   Which of the following will not accelerate deflation?
     A.   established vegetation
     B.   construction
     C.   motor vehicle tracks
     D.   plowing
Answer = A (page 352)

17.   How does desert pavement form?
     A.   by intense chemical weathering
     B.   by stream erosion
     C.   by intense mechanical weathering
     D.   by wind erosion
Answer = D (page 352)

18.   Yardangs are _____.
     A.   sand dunes arranged in long, parallel rows
     B.   eroded bedrock with long, parallel grooves
     C.   long, parallel ridges eroded by dust
     D.   cobbles with long, parallel pits
Answer = C (page 353)

19.   Which of the following promote the development of sand dunes?
     A.   strong winds
     B.   a supply loose sand
     C.   dry climate
     D.   all of these
Answer = D (page 354)

20. Sand will accumulate _____.
     A.   on the leeside (downwind) of a boulder
     B.   on the windward side (upwind) of a boulder
     C.   equally on the leeside and the windward side of a boulder
     D.   neither on the leeside nor on the windward side of a boulder
Answer = A (page 355)

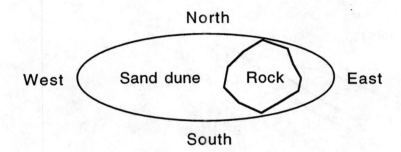

21. Given the relative positions of the sand dune and rock shown here, the wind direction is predominantly _____.
     A.   north to south
     B.   south to north
     C.   east to west
     D.   west to east
Answer = C (page 355)

22. The steeper, downwind side of a sand dune is called a _____.
     A.   sand drift
     B.   ventifact
     C.   slip face
     D.   streamline
Answer = C (page 355)

23. How high are the highest sand dunes in the world?
     A.   2.5 meters
     B.   25 meters
     C.   250 meters
     D.   2.5 kilometers
Answer = C (page 356)

24. A cross bed preserved in an eolian sandstone represents _____.
     A.   the top of a former sand dune
     B.   the bottom of a former sand dune
     C.   the upwind side of a former sand dune
     D.   the downwind side of a former sand dune
Answer = D (page 356)

25.   The cross beds exposed in a sandstone dip to the west. During deposition of the sand, the prevailing winds were probably from the _____.
    A.   east
    B.   west
    C.   north
    D.   south
Answer = A (page 356)

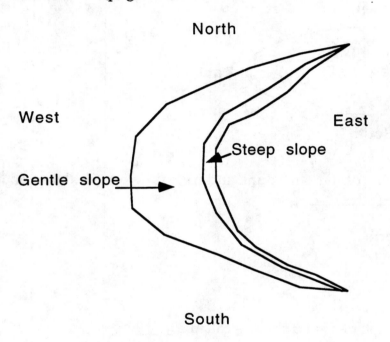

26.   This sand dune is a _____ dune.
    A.   barchan
    B.   transverse
    C.   blowout
    D.   linear
Answer = A (page 357)

27.   The wind direction which would create this dune is _____.
    A.   north to south
    B.   south to north
    C.   east to west
    D.   west to east
Answer = D (page 357)

28.   In a barchan dune, the points of the crescent point _____ and the slip face is the _____ downwind curve of this dune.
        A.   upwind ... concave
        B.   upwind ... convex
        C.   downwind ... concave
        D.   downwind ... convex
    Answer = C (page 357)

29.   Sand dunes behind beaches are typically _____.
        A.   transverse dunes
        B.   barchans
        C.   linear dunes
        D.   draas
    Answer = A (page 358)

30.   Large dune fields, or "seas of sand," are called _____.
        A.   draas
        B.   barchans
        C.   loess
        D.   ergs
    Answer = D (page 358)

31.   During erosion, loess tends to break off _____.
        A.   in horizontal layers
        B.   in vertical sheets
        C.   along irregular cracks
        D.   slowly, grain-by-grain
    Answer = B (page 358)

32.   Which of the following is not related to wind erosion?
        A.   loess
        B.   ventifact
        C.   deflation
        D.   blowout
    Answer = A (page 358)

33.   Long sand ridges that are more or less parallel to the prevailing wind are called _____.
        A.   barchan dunes
        B.   linear dunes
        C.   transverse dunes
        D.   blowouts
    Answer = B (page 358)

34.   The loess deposited in the upper Mississippi Valley was derived primarily from
         _____.
                    A.   volcanic dust
                    B.   desert regions lying to the west
                    C.   glacial deposits
                    D.   coastal sand dunes
         Answer = C (page 359)

35.   The loess deposits of the Mississippi Valley are due to _____.
                    A.   glacial activity and wind in the Pleistocene epoch
                    B.   erosion of the Appalachian Mountains in the Oligocene epoch
                    C.   drying of large lakes in the Miocene epoch
                    D.   desert conditions in the Paleocene epoch
         Answer = A (page 359)

36.   Why does volcanic dust tend to travel further than wind-blown dust derived from
         the continents?
                    A.   because volcanic dust is erupted high into the atmosphere
                    B.   because volcanic dust is less dense than dust derived from the conti-
                          nents
                    C.   because volcanic dust is composed of quartz whereas dust derived
                          from the continents is composed of feldspar
                    D.   all of these
         Answer = A (page 359)

37.   Which of the following statements is false?
                    A.   Deserts can occur under virtually stationary areas of low atmospheric
                          pressure.
                    B.   Deserts can occur so far inland that air precipitates all moisture before
                          reaching the desert area.
                    C.   Deserts can occur where air is so cold only small amounts of moisture
                          can precipitate.
                    D.   Deserts can occur where moisture-bearing winds are blocked by moun-
                          tain ranges.
         Answer = A (pages 359, 360)

38.   The Great Basin and Mojave deserts of western North America exist primarily
         _____.
                    A.   because they lie near the equator
                    B.   because they lie thousands of kilometers from the ocean
                    C.   because they lie in the rain shadow of coastal mountains
                    D.   all of these
         Answer = C (page 360)

39. Through geologic time, the interior of Australia has changed from a moist, humid climate to a desert because _____.
    A.   plate collisions have built mountains near the coastline which block moisture-bearing winds from the continent's interior
    B.   Australia has moved northward into an arid, subtropical zone
    C.   Australia has moved southward out of the path of the more humid trade winds
    D.   the continent has grown through volcanic activity, making the interior too far away from moisture-bearing winds
    Answer = B (page 360)

40. Radar imaging from the space shuttle Endeavor has provided evidence that the Sahara Desert _____.
    A.   has always been an arid environment
    B.   was once a tropical region, with many different types of vegetation
    C.   once had an extensive system of river channels, which are now dry
    D.   has sand dunes in excess of 300 m in height
    Answer = C (page 361)

41. The transformation of semi-arid regions into deserts is called _____.
    A.   deflation
    B.   deforestation
    C.   delamination
    D.   desertification
    Answer = D (page 361)

42. Much of a desert's surface consists of sand, gravel, and rock rubble because _____.
    A.   clay minerals form slowly in a desert environment
    B.   wind blows away clay and soil before it can accumulate to great thickness
    C.   vegetation is sparse and cannot prevent erosion of soil
    D.   all of the above
    Answer = D (page 361)

43. What minerals are responsible for the orange-brown colors of weathered surfaces in the desert?
    A.   quartz and feldspar
    B.   carbonates
    C.   iron oxides
    D.   all of these
    Answer = C (page 361)

44.   Desert varnish is composed of _____.
      A.   clay minerals, iron oxides, and zinc
      B.   manganese, zinc, and clay minerals
      C.   clay minerals, iron oxides, and manganese
      D.   iron oxides, manganese, and silica
Answer = C (page 361)

45.   Approximately how much of the world's deserts are covered by sand?
      A.   5%
      B.   20%
      C.   50%
      D.   100%
Answer = B (page 363)

46.   Which of the following would you least expect to find in a desert?
      A.   playa lakes
      B.   rounded, soil-covered hills
      C.   steep river valleys and gorges
      D.   dune fields
Answer = B (page 364)

47.   A wadi is a _____.
      A.   shallow lake
      B.   playa with salt sediments
      C.   dry wash
      D.   flood deposit
Answer = C (page 364)

48.   In the desert, a broad, gently-sloping platform of bedrock that is left behind as a mountain front is eroded is called a(n) _____.
      A.   pediment
      B.   alluvial fan
      C.   mesa
      D.   erg
Answer = A (page 365)

49.   Which of the following is not a wind deposit?
      A.   pediment
      B.   sand dune
      C.   loess
      D.   all of these are wind deposits
Answer = A (page 365)

50.   A mesa is a _____.
   A.   flood deposit in a small dry wash
   B.   flat plateau surrounded by steep cliffs
   C.   mountain showing horizontal sedimentary layering
   D.   pediment covered with an alluvial fan
  Answer = B (page 366)

15

Glaciers: The Work of Ice

1. Widespread glacial erosion and sedimentation do not affect _____.
    A. water discharge and sediment loads of major river systems
    B. desertification of semi-arid lands
    C. quantity of sediment delivered to the oceans
    D. erosion and sedimentation in coastal areas and on shallow continental shelves.
    Answer = B (page 372)

2. Glacial ice forms by _____ of snow.
    A. burial and metamorphism
    B. melting and refreezing
    C. erosion and deposition
    D. precipitation and melting
    Answer = A (page 372)

3. Large masses of ice on land that show evidence of movement are called _____.
    A. ice packs
    B. glaciers
    C. icebergs
    D. all of these
    Answer = B (page 373)

4. Valley glaciers are also known as _____ glaciers.
    A. lowland
    B. upland
    C. alpine
    D. gorge
    Answer = C (page 373)

5.   Ice covers what percentage of the Antarctic continent?
          A.   50 %
          B.   75%
          C.   90%
          D.   100%
      Answer = C (page 374)

6.   High latitudes are cold because _____.
          A.   there are fewer hours of sunlight than at the equator
          B.   there is more snow than at the equator
          C.   the angle between the Sun's rays and the Earth's surface is different
                than at the equator
          D.   the Earth's poles are further from the Sun than the Earth's equator
      Answer = C (page 375)

7.   A deposit of snow contains approximately 90% air. Glacial ice, which forms from
      snow, contains _____ air.
          A.   approximately 90%
          B.   approximately 50%
          C.   approximately 25%
          D.   less than 20%
      Answer = D (page 375)

8.   At the equator, at what altitude does the snow line lie?
          A.   less than 1000 meters
          B.   about 2500 meters
          C.   about 5000 meters
          D.   about 7500 meters
      Answer = C (page 375)

9.   Which of the following lists is written in order of increasing ice "metamorphism"?
          A.   snow --> granular ice --> firn --> glacial ice
          B.   snow --> firn --> glacial ice --> granular ice
          C.   snow --> firn --> granular ice --> glacial ice
          D.   snow --> granular ice --> glacial ice --> firn
      Answer = A (page 375)

10.   Which of the following refers to the total amount of ice lost from a glacier each year?
          A.   meltage
          B.   accumulation
          C.   sublimation
          D.   ablation
      Answer = D (page 376)

11.   The transformation of ice to gaseous water vapor is called _____.
          A.   melting
          B.   sublimation
          C.   boiling
          D.   none of these
        Answer = B (page 376)

12.   Which of the following processes is not a form of glacial ablation?
          A.   melting
          B.   sublimation
          C.   calving
          D.   All of these are forms of glacial ablation.
        Answer = D (page 376)

13.   If accumulation exceeds ablation, then _____.
          A.   the glacial ice will flow downhill but the end of the glacier will move
                  uphill
          B.   the glacial ice will flow downhill and the end of the glacier will move
                  downhill
          C.   the glacial ice will flow uphill and the end of the glacier will move
                  uphill
          D.   the glacial ice will flow uphill but the end of the glacier will move
                  downhill
        Answer = B (page 377)

14.   Cold, dry glaciers move _____.
          A.   mostly by plastic flow
          B.   mostly by basal slip
          C.   by roughly equal amounts of plastic flow and basal slip
          D.   by neither plastic flow nor basal slip
        Answer = A (page 378)

15.   Crevasses form because _____.
          A.   the glacial surface partially melts, leaving holes and cracks
          B.   glacial meltwater erodes small valleys as glacial rivers flow
          C.   low confining pressure at the surface causes cracks as the ice flows
          D.   a glacial calving process has not been completed
        Answer = C (page 379)

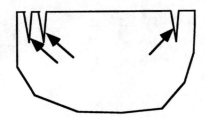

16.   In this cross-section of a glacier, the arrows point to _____.
         A.   ablation surfaces
         B.   blocks of bedrock
         C.   crevasses
         D.   ice streams
       Answer = C (page 379)

17.   An ice stream is _____.
         A.   a river of meltwater running on top of a glacier
         B.   a river of meltwater running beneath a glacier
         C.   an area of mostly solid ice which moves faster than the surrounding
               glacial ice
         D.   a path of arctic oceanic water circulation north of 60°N latitude
       Answer = C (page 380)

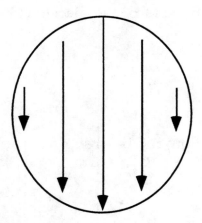

18.   If the relative length of the arrows represents relative speed of flow, the flow pattern
       shown here is most likely that of _____.
         A.   a valley glacier
         B.   a continental glacier
         C.   either a valley or a continental glacier
         D.   neither a valley nor a continental glacier
       Answer = A (page 380)

19. What is the term for a sudden period of fast movement of a valley glacier?
    A. sublimation
    B. surge
    C. firn
    D. drift
    Answer = B (page 381)

20. In a continental glacier, ice flows _____.
    A. from the edges to the center
    B. from the center to the edges
    C. from north to south
    D. from south to north
    Answer = B (page 381)

21. Which of the following can be used to determine the direction a continental glacier moved?
    A. striations
    B. roches moutonées
    C. drumlins
    D. all of these
    Answer = D (pages 381, 382, 386)

22. What is the term for an amphitheater-like hollow that forms at the head of a glacier?
    A. kettle
    B. moraine
    C. cirque
    D. drumlin
    Answer = C (page 382)

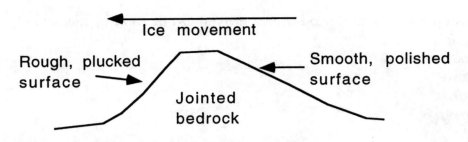

23.   This cross-sectional diagram depicts a _____.
         A.   roche moutonée
         B.   drumlin
         C.   kame
         D.   esker
Answer = A (page 382)

24.   When two cirques at the heads of adjacent valleys meet at the moutaintop, they produce a sharp, jagged crest called _____.
         A.   an arete
         B.   a fjord
         C.   a drumlin
         D.   a moraine
Answer = A (page 382)

25.   A hanging valley is _____.
         A.   a valley with more than three tributaries feeding into it
         B.   a valley with many loose rocks in its walls ready for a landslide
         C.   a valley with its floor high above the main valley floor
         D.   a valley feeding into many fjords
Answer = C (page 383)

26.   Erratics are _____.
         A.   cross beds which do not match the overall outwash cross-bedding
                scheme
         B.   large boulders deposited randomly by glaciers
         C.   valleys with glacial striations that oppose the striations in adjacent val-
                leys
         D.   conflicting dates of glaciation obtained by dating organic material
                found in glaciers
Answer = B (page 384)

27.   Which of the following agents of erosion deposits the most poorly sorted sediment?
      A.   wind
      B.   ice
      C.   streams
      D.   ocean currents
Answer = B (page 384)

28.   What feature forms where two lateral moraines merge?
      A.   end moraine
      B.   medial moraine
      C.   ground moraine
      D.   eskers
Answer = B (page 385)

29.   Which of the following is a glacial deposit and not a glacial erosional feature?
      A.   a cirque
      B.   a moraine
      C.   an arete
      D.   all of these are glacial erosional features
Answer = B (page 385)

30.   Large, streamlined hills of till and bedrock that form in some areas subjected to continental glaciation are called _____.
      A.   aretes
      B.   kames
      C.   eskers
      D.   drumlins
Answer = D (page 386)

31.   Varve deposits are formed when _____ is deposited in the summer and _____ is deposited in the winter.
      A.   coarse silt ... fine clay
      B.   gravel ... sand
      C.   fine clay ... sand
      D.   coarse silt ... gravel
Answer = A (page 387)

32.   Which of the following would be best for a sand and gravel pit?
      A.   a drumlin
      B.   a kame
      C.   a varve
      D.   an arete
Answer = B (page 387)

33. Which of the following are not water-laid deposits associated with glaciers?
    A. moraines
    B. kames
    C. eskers
    D. varves
    Answer = A (pages 387, 388)

34. Kettles are formed by _____.
    A. meltwater streams running through glacial tunnels
    B. large blocks of ice left by a glacier
    C. seasonal deposition of coarse and fine sediment
    D. melting permafrost
    Answer = B (page 388)

35. How much of the Earth's land surface is permanently frozen?
    A. approximately 1%
    B. approximately 5%
    C. approximately 10%
    D. approximately 25%
    Answer = D (page 388)

36. Long, narrow, winding ridges of sand and gravel found in the middle of ground moraines are called _____.
    A. aretes
    B. eskers
    C. kames
    D. drumlins
    Answer = B (page 388)

37. Who deduced that large continental glaciers once covered much of the Earth's land surface?
    A. Charles Darwin
    B. James Hutton
    C. Louis Agassiz
    D. Charles Lyell
    Answer = C (page 389)

38. If all of the world's ice were to melt, _____.
    A. sea level would not change
    B. sea level would rise about 6 meters
    C. sea level would rise about 60 meters
    D. sea level would rise about 600 meters
    Answer = C (page 390)

39.   The Earth is expected to go back into a period of glaciation over the next _____ years.
        A.   100
        B.   1,000
        C.   10,000
        D.   100,000
Answer = C (page 390)

40.   The age of the last glacial period is dated by _____.
        A.   carbon-14 dating of logs in the glacial drift
        B.   carbon-14 dating of wooly mammoths in glacial ice
        C.   carbon-14 dating of plant leaves contained in the glacial drift
        D.   carbon-14 dating of marine organisms contained in glacial ice
Answer = A (page 391)

41.   The recent ice ages occurred during the _____ Epoch.
        A.   Pliocene
        B.   Paleozoic
        C.   Pleistocene
        D.   Jurassic
Answer = C (page 391)

42.   How many distinct glaciations affected North America during the Pleistocene Epoch?
        A.   one
        B.   two
        C.   three
        D.   four
Answer = D (page 391)

43.   During the peak of the last ice age, sea level was about _____ than sea level today.
        A.   1 meters lower
        B.   10 meters lower
        C.   100 meters lower
        D.   1000 meters lower
Answer = C (page 392)

44.   Approximately how thick was the ice cap that covered northern North America during the last ice age?
        A.   300 meters
        B.   1,000 meters
        C.   3,000 meters
        D.   10,000 meters
Answer = C (page 392)

45.   The polar ice caps of today began forming _____ years ago.
         A.   10,000 (ten thousand)
         B.   100,000 (one hundred thousand)
         C.   1,000,000 (one million)
         D.   10,000,000 (ten million)
       Answer = D (page 393)

46.   The degree of ellipticity of Earth's orbit around the sun is called _____.
         A.   ellipticality
         B.   eccentricity
         C.   ovality
         D.   precession
       Answer = B (page 394)

47.   Precession describes _____.
         A.   the movement of ice sheets away from the poles
         B.   the movement of glacial ice down through valleys
         C.   the wobble of the Earth's axis of rotation
         D.   the changes in climate as the Earth warms or cools
       Answer = C (page 394)

48.   Carbon dioxide in the atmosphere _____.
         A.   is low during glacial periods
         B.   is high during glacial periods
         C.   is constant between glacial and interglacial periods
         D.   is inconsistent with respect to glacial and interglacial periods
       Answer = A (page 395)

49.   The "Little Ice Age" of 1400–1650 A.D., is characterized as such in part because
       _____.
         A.   the Baltic Sea froze over
         B.   the Red Sea froze over
         C.   the Dead Sea froze over
         D.   none of the above
       Answer = A (page 395)

50.   Which of the following is not one of the possible explanations for previous ice ages?
         A.   large land masses (continents) positioned at the poles
         B.   periodic changes in the Earth's orbit
         C.   increases in the amount of carbon dioxide in the atmosphere
         D.   all of these are possible explanations for the ice ages
       Answer = C (page 398)

16
Landscape Evolution

1.   Erosion and sedimentation are powered by _____.
          A.   tectonics
          B.   the sun
          C.   radioactivity
          D.   tidal forces
     Answer = A (page 401)

2.   Accurate maps of the Earth's surface are necessary for _____.
          A.   hydrologists
          B.   glaciologists
          C.   ecologists
          D.   all of the above
     Answer = D (page 402)

3.   On a topographic map, what type of lines connect points of equal elevation?
          A.   contours
          B.   isograds
          C.   contacts
          D.   horizons
     Answer = A (page 402)

4.   On a topographic map, the more closely spaced the contour lines, _____.
          A.   the higher the mountain
          B.   the lower the mountain
          C.   the flatter the slope
          D.   the steeper the slope
     Answer = D (page 402)

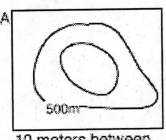
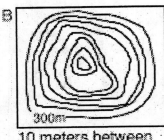
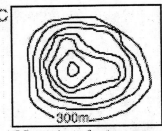

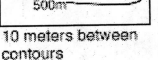

10 meters between contours   10 meters between contours   20 meters between contours

5.   Which of the contour diagrams depicts the landform with the highest elevation?
   A.   diagram A
   B.   diagram B
   C.   diagram C
   D.   cannot tell from information given
Answer = A (page 402)

6.   Which of the contour diagrams depicts the landform with the lowest relief?
   A.   diagram A
   B.   diagram B
   C.   diagram C
   D.   cannot tell from information given
Answer = A (page 403)

7.   The relief of an area is best described as _____.
   A.   the elevation of the highest point in the area minus the elevation of sea level
   B.   the elevation of the lowest point in the area minus the elevation of sea level
   C.   the elevation of the highest point in the area minus the elevation of the lowest point in the area
   D.   the elevation of the lowest point in the area minus the elevation of the highest point in the area
Answer = C (page 403)

8.   On a global scale, the global relief of the land surface is approximately _____.
   A.   3 kilometers
   B.   9 kilometers
   C.   30 kilometers
   D.   90 kilometers
Answer = B (page 404)

9.   The shapes of erosion and sedimentation are called _____.
     A.   landforms
     B.   relief
     C.   contours
     D.   bedding
    Answer = A (page 404)

10.   What is the approximate relief present in the Appalachian Mountains?
     A.   about 200 meters
     B.   about 800 meters
     C.   about 3200 meters
     D.   about 10,000 meters
    Answer = B (page 404)

11.   Which of the following regions has the lowest relief?
     A.   the Appalachian Mountains
     B.   the midwestern plains
     C.   the Rocky Mountains
     D.   all of these have approximately the same, low relief
    Answer = B (page 404)

12.   Which of the following statements about mountains is false?
     A.   Mountains are distinguished from hills only by size and by custom.
     B.   Elevations that would be considered mountains in one terrain could be hills in another terrain.
     C.   Volcanoes which rise as only a single peak are not considered mountains.
     D.   Landforms more than several hundred meters above their surroundings are generally considered to be mountains.
    Answer = C (page 405)

13.   The youngest mountains in the world are the _____.
     A.   Appalachians
     B.   Rockies
     C.   Andes
     D.   Himalayas
    Answer = D (page 405)

14.   The steepest slopes are generally found in _____.
     A.   mountains at low elevations with low relief
     B.   mountains at high elevations with low relief
     C.   mountains at low elevations with high relief
     D.   mountains at high elevations with high relief
    Answer = D (page 405)

15.   Most plateaus have elevations of _____.
    A.   less than 1000 meters
    B.   less than 3000 meters
    C.   less than 5000 meters
    D.   more than 5000 meters
Answer = B (page 405)

16.   What is the name for a small landform with a flat elevation and steep slopes on all sides?
    A.   dome
    B.   mesa
    C.   cuesta
    D.   plateau
Answer = B (page 405)

17.   What is the average elevation of the Tibetan Plateau?
    A.   200 meters
    B.   500 meters
    C.   2,000 meters
    D.   5,000 meters
Answer = D (page 405)

18.   The Altiplano of Bolivia has an average elevation of _____.
    A.   540 m
    B.   1100 m
    C.   3600 m
    D.   5300 m
Answer = C (page 405)

19.   Many plateaus are flat because _____.
    A.   erosion was constant over a broad region
    B.   shallow seas eroded the surface
    C.   they are underlain by undeformed sedimentary rocks or layers of lava flows
    D.   none of these
Answer = C (page 405)

20.   A mesa is an example of _____.
    A.   a depositional landform
    B.   an erosional landform
    C.   both a depositional and an erosional landform
    D.   neither a depositional nor a erosional landform
Answer = B (page 405)

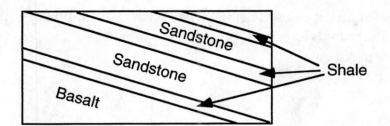

21. If the rock package represented in this cross section were to be eroded, what type of landform would it most likely create?
   A. cuesta
   B. hogback
   C. mesa
   D. mountain

   Answer = A (page 406)

22. Which of the rock units in this diagram is most easily eroded?
   A. sandstone
   B. shale
   C. basalt
   D. cannot be determined from the information given

   Answer = B (page 406)

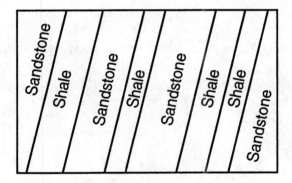

23. If the rock package represented in this cross-sectional diagram were to be eroded, what type of landform would it most likely create?
   A. cuesta
   B. hogback
   C. mesa
   D. mountain

   Answer = B (page 406)

24. What is the geologic term for an asymmetric ridge that consists of shallowly dipping tilted and eroded sedimentary beds?
    A.   cuesta
    B.   yardang
    C.   hogback
    D.   mesa
    Answer = A (page 406)

25. What is the geologic term for narrow ridges formed by layers of erosion-resistant sedimentary rocks which are tilted to nearly vertical?
    A.   cuestas
    B.   yardangs
    C.   hogbacks
    D.   mesas
    Answer = C (page 406)

26. Which of the following types of river valleys indicates the most advanced erosion state?
    A.   an open valley in hills
    B.   a broad, flat valley in lowlands
    C.   a narrow mountain canyon
    D.   none of these indicate an advanced erosion state
    Answer = B (page 408)

27. A deeply gullied topography that results from the rapid erosion of easily eroded shales and clays is called _____.
    A.   a cuesta topography
    B.   badlands
    C.   a karst topography
    D.   none of these
    Answer = B (page 408)

28. Badlands are formed from _____ rocks.
    A.   sedimentary
    B.   igneous
    C.   metamorphic
    D.   any of the above is equally likely
    Answer = A (page 408)

29. Badlands are an example of _____.
   A. a primarily depositional landform
   B. a primarily erosional landform
   C. both depositional and erosional landforms
   D. neither depositional nor erosional landforms
   Answer = B (page 408)

30. Which of the following rivers occurs in a rift valley?
   A. the Mississippi River
   B. the Sacramento River
   C. the Amazon River
   D. the River Jordan
   Answer = D (page 408)

31. The Appalachian Mountains are relatively low compared to the Rocky Mountains indicating that _____.
   A. the Appalachians are probably younger than the Rockies
   B. the Appalachians are probably older than the Rockies
   C. the Appalachians and the Rockies are probably the same age
   D. erosion is more intense in the Appalachians
   Answer = B (page 410)

32. A process in which one action produces an effect that slows down the original action and stabilizes it at a slower rate is called a _____.
   A. positive-feedback process
   B. no-feedback process
   C. negative-feedback process
   D. none of these
   Answer = C (page 410)

33. If the average tectonic uplift rate of a mountain range increases, the average erosion rate will generally _____.
   A. increase
   B. decrease
   C. remain the same
   D. decrease first, then increase
   Answer = A (page 410)

34. Erosion is _____ when uplift is _____.
   A. fastest ... fastest
   B. fastest ... slowest
   C. slowest ... fastest
   D. slowest ... moderate
   Answer = A (page 410)

35.   Climbing higher in elevation is equivalent to _____.
      A.   traveling to lower latitudes
      B.   traveling to a desert region
      C.   traveling to higher latitudes
      D.   traveling to a rain forest
Answer = C (page 411)

36.   In which of the following climates would you expect to find rugged topography?
      A.   a tropical rain forest
      B.   a temperate climate with moderate rainfall
      C.   a desert
      D.   all of these are equally likely to have rugged topography
Answer = C (page 411)

37.   Chemical weathering is likely to be least important _____.
      A.   in a humid, tropical climate
      B.   at high latitudes near the poles
      C.   in an area of heavy rainfall
      D.   at the equator
Answer = B (page 411)

38.   What mountain chain runs along the east coast of North America?
      A.   the Appalachians
      B.   the Sierra Nevada
      C.   the Cascades
      D.   the Rockies
Answer = A (page 412)

39.   Along the west coast of North America, the North American plate is in contact with which of the following tectonic plates?
      A.   the Gorda plate
      B.   the Juan de Fuca plate
      C.   the Pacific plate
      D.   all of these
Answer = D (page 412)

40.   Which of the following rivers flows westward from the Appalachian Mountains?
      A.   Mississippi River
      B.   Missouri River
      C.   Ohio River
      D.   Colorado River
Answer = C (page 412)

41.   Most of the plains of the central United States and southern Canada are floored by horizontal sedimentary rocks of _____ age.
          A.   Cenozoic
          B.   Mesozoic
          C.   Paleozoic
          D.   Precambrian
       Answer = C (page 412)

42.   What geographic region lies between the Rocky Mountains and the Sierra Nevada?
          A.   the Great Plains
          B.   the Basin and Range province
          C.   the Cascades
          D.   the Coast Ranges
       Answer = B (page 412)

43.   In which region of North America does active plate tectonics affect the landscape?
          A.   along the east coast
          B.   in the central part of the continent
          C.   along the west coast
          D.   A and C
       Answer = C (page 412)

44.   The low coastal plains of the central and southern Atlantic coast are comprised primarily of which rock type?
          A.   sedimentary rock
          B.   metamorphic rock
          C.   volcanic rock
          D.   plutonic rock
       Answer = A (page 412)

45.   Which of the following regions is the oldest geologically?
          A.   the east coast of North America
          B.   the central plains of North America
          C.   the west coast of North America
          D.   the northern part of North America
       Answer = D (page 412)

46.   The progression from young, high, rugged mountains to older, worn-down, rounded hills is called the _____.
          A.   rock cycle
          B.   Wilson cycle
          C.   cycle of erosion
          D.   hydrologic cycle
       Answer = C (page 414)

47.   The "cycle of erosion" was proposed by _____.
       A.   William Morris Davis
       B.   James Hutton
       C.   Charles Darwin
       D.   Alfred Wegener
       Answer = A (page 414)

48.   The flat surfaces of _____ were thought to be proof that old, level plains result from continued erosion.
       A.   faults
       B.   joints
       C.   crevasses
       D.   unconformities
       Answer = D (page 414)

49.   With what type of convergent plate boundary are the highest mountains on Earth associated?
       A.   ocean-ocean convergence
       B.   ocean-continent convergence
       C.   continent-continent convergence
       D.   none of these
       Answer = C (page 415)

50.   The rate of uplift of a land surface varies with the speed of _____.
       A.   plate convergence
       B.   earthquakes
       C.   seismic wave travel time
       D.   sediment accumulation
       Answer = A (page 415)

17

The Oceans

1.   Oceans cover approximately _____ percent of the Earth's surface.
            A.   30
            B.   50
            C.   70
            D.   90
        Answer = C (page 419)

2.   Ripples in the ocean grow to full-sized waves when the wind reaches a speed of about _____ per hour.
            A.   3 meters
            B.   30 meters
            C.   3 kilometers
            D.   30 kilometers
        Answer = D (page 421)

3.   The height of an ocean wave increases as _____.
            A.   the wind speed increases
            B.   the wind blows for longer times
            C.   the distance over which the wind blows over the water increases
            D.   all of these
        Answer = D (page 421)

4.   Waves cause small particles floating on the surface to move in _____.
            A.   horizontal elliptical orbits
            B.   vertical elliptical orbits
            C.   horizontal circular orbits
            D.   vertical circular orbits
        Answer = D (page 421)

5.   The distance between two wave crests is called the _____.
          A.   wavelength
          B.   wave height
          C.   throw
          D.   period
       Answer = A (page 422)

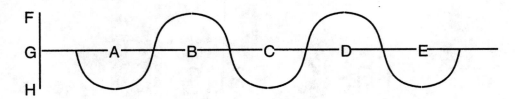

6.   In this diagram, the wavelength is the distance _____.
          A.   A to B
          B.   A to C
          C.   A to D
          D.   A to E
       Answer = B (page 422)

7.   In this diagram, the distance F to H is called the _____.
          A.   crest height
          B.   trough height
          C.   wave height
          D.   total displacement
       Answer = C (page 422)

8.   Which of the following equations correctly relates the velocity (V) of a wave to the
       wavelength (L) and period (T)?
          A. $V = L \times T$
          B. $V = L/T$
          C. $V = L^2 \times T$
          D. $V = T/L$
       Answer = B (page 422)

9.   The orbital motion of water particles due to surface waves stops at a depth
       _____.
          A.   of about twice the wavelength
          B.   equal to the wavelength
          C.   of about one-half the wavelength
          D.   of about one-tenth the wavelength
       Answer = C (page 422)

10.   When a wave gets close to shore, water particles near the bottom move in _____ orbits.
        A.   circular
        B.   elliptical
        C.   irregular
        D.   rectangular
Answer = B (page 422)

11.   Waves run up onto the beach forming a _____, and run back down as _____.
        A.   swash ... backwash
        B.   wave ... swash
        C.   front swash ... retrowash
        D.   wave ... backwash
Answer = A (page 423)

12.   As waves approach a beach, the rows of waves gradually bend to a direction more parallel to the shore. This change in direction is called _____.
        A.   longshore drift
        B.   swash
        C.   tidal surge
        D.   wave refraction
Answer = D (page 424)

13.   _____ refers to the movement of sand grains along the beach.
        A.   Wave refraction
        B.   Turbidity current
        C.   Longshore drift
        D.   Ebb tide
Answer = C (page 424)

14.   The zigzag motion which carries sand grains along a beach is known as _____.
        A.   turbidity
        B.   longshore drift
        C.   refraction
        D.   tides
Answer = B (page 424)

15.   How many high tides are there in a day?
        A.   1
        B.   2
        C.   3
        D.   4
Answer = B (page 426)

16.   What causes the tides?
     A.   wind
     B.   seismicity
     C.   ocean currents
     D.   gravity
Answer = D (page 426)

17.   Spring tides occur _____.
     A.   once a year
     B.   once a month during March, April, and May only
     C.   once a month
     D.   once every two weeks
Answer = D (page 427)

18.   A tsunami is _____.
     A.   a tidal surge caused by a storm
     B.   a large wave caused by an undersea event
     C.   an earthquake which causes a large wave
     D.   an undersea landslide
Answer = B (page 427)

19.   The muddy or sandy areas that are exposed during low tide, but are flooded at high tide are called _____.
     A.   estuaries
     B.   tidal flats
     C.   surf zones
     D.   passive margins
Answer = B (page 428)

20.   The surf zone lies in the _____ part of the beach.
     A.   offshore
     B.   foreshore
     C.   backshore
     D.   all of these
Answer = B (page 428)

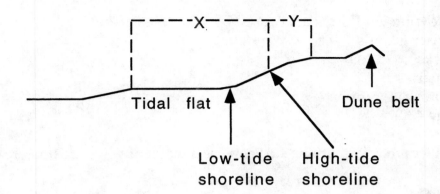

Tidal flat

Low-tide shoreline

High-tide shoreline

Dune belt

21.   The area labeled "X" is called the _____
        A.   foreshore
        B.   surf zone
        C.   swash zone
        D.   backshore
Answer = B (page 429)

22.   The area labeled "Y" is called the _____
        A.   foreshore
        B.   surf zone
        C.   swash zone
        D.   backshore
Answer = C (page 429)

23.   Together, the areas labeled "X" and "Y" are called the _____
        A.   foreshore
        B.   surf zone
        C.   swash zone
        D.   backshore
Answer = A (page 429)

24.   If sand input is greater than sand output in a beach's sand budget, the beach will
       _____.
        A.   be long, wide, and sandy
        B.   have large sand dunes
        C.   have large tidal flats
        D.   have high cliffs
Answer = A (page 430)

25.   Which of the following landforms are caused by wave erosion?
        A.   barrier islands
        B.   stacks
        C.   estuaries
        D.   submarine canyons
        Answer = B (page 432)

26.   Planar rocky surfaces that form in the surf zone as a result of wave erosion are called

        _____.
        A.   stacks
        B.   backshores
        C.   wave-cut terraces
        D.   barrier islands
        Answer = C (page 432)

27.   What do we call long offshore sandbars?
        A.   barrier islands
        B.   estuaries
        C.   stacks
        D.   none of these
        Answer = A (page 433)

28.   Sea-level changes are detected by geologic studies of _____.
        A.   beaches
        B.   cliffs along sea shores
        C.   structures built along beaches
        D.   wave-cut terraces
        Answer = D (page 434)

29.   A coastal body of water connected to the ocean and supplied with fresh water from
        a river is a(n) _____.
        A.   river
        B.   atoll
        C.   spit
        D.   estuary
        Answer = D (page 435)

Topographic profile of the Atlantic Ocean from New England to the mid-ocean ridge.

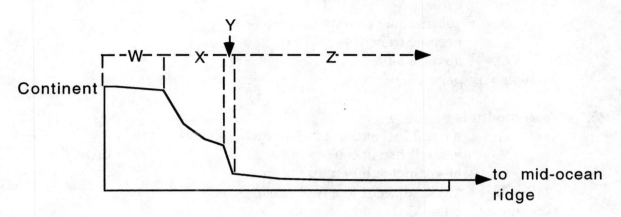

30.   The area labeled "W" is called the _____.
        A.   abyssal plain
        B.   continental slope
        C.   shoreline
        D.   continental shelf
Answer = D (page 436)

31.   The area labeled "X" is called the _____.
        A.   ocean trough
        B.   continental slope
        C.   continental rise
        D.   ocean trench
Answer = B (page 436)

32.   The area labeled "Y" is called the _____.
        A.   ocean trough
        B.   continental slope
        C.   continental rise
        D.   ocean trench
Answer = C (page 436)

33.   The area labeled "Z" is called the _____.
        A.   abyssal plain
        B.   oceanic slope
        C.   continental rise
        D.   oceanic shelf
Answer = A (page 436)

34.   Deep valleys eroded into the continental slope and shelf are called _____.
          A.   estuaries
          B.   submarine canyons
          C.   oceanic trenches
          D.   abyssal valleys
       Answer = B (page 437)

35.   A seamount is _____.
          A.   a coral reef above a submerged volcano
          B.   a small rise in the sea floor
          C.   an extinct submerged volcano
          D.   a hill adjacent to a mid-ocean ridge
       Answer = C (page 438)

36.   A seamount that rises 1500 m from the seafloor will raise the ocean surface by
       _____ meters.
          A.   0.15 meters
          B.   1.5 meters
          C.   15 meters
          D.   150 meters
       Answer = B (page 439)

37.   Where is the ocean floor deepest?
          A.   in rift valleys
          B.   in the abyssal plain
          C.   in submarine canyons
          D.   in oceanic trenches
       Answer = D (page 440)

38.   Which of the following is an active continental margin?
          A.   the east coast of North America
          B.   the west coast of South America
          C.   the east coast of Africa
          D.   all of these
       Answer = B (page 440)

39.   What type of currents erode and deposit fine-grained sediments on the continental
       slope and rise?
          A.   tidal currents
          B.   turbidity currents
          C.   longshore currents
          D.   river currents
       Answer = B (page 441)

40. Turbidity currents are most likely to be found _____.
   A. where rivers empty into oceans
   B. on the continental shelf
   C. on the continental slope
   D. on the abyssal plains away from the continental rise
   Answer = C (page 444)

41. Large fan-shaped deposits of fine-grained sediments that accumulate on the continental rise are called _____.
   A. submarine fans
   B. atolls
   C. alluvial fans
   D. spits
   Answer = A (page 444)

42. Superheated water approximately _____ spouts from the hot springs on the East Pacific Rise.
   A. 4°C
   B. 40°C
   C. 400°C
   D. 4000°C
   Answer = C (page 445)

43. A volume of seawater equal to the volume of the ocean cycles through submarine hydrothermal systems roughly every _____.
   A. year
   B. five thousand years
   C. ten million years
   D. billion years
   Answer = C (page 445)

44. Most volcanic activity on the seafloor takes place on _____.
   A. continental shelves
   B. abyssal plains
   C. continental rises
   D. mid-ocean ridges
   Answer = D (page 446)

45. Black smokers are full of _____.
   A. carbon dioxide and metals
   B. dissolved hydrogen sulfide and metals
   C. oxygen and metals
   D. nitrogen and metals
   Answer = B (page 446)

46.   What is the maximum water depth of living coral reefs?
    A.   about 20 meters
    B.   about 100 meters
    C.   about 500 meters
    D.   about 2 kilometers
Answer = A (page 448)

47.   Pelagic sediments consist of _____.
    A.   reddish-brown clays derived from the continents
    B.   foraminiferal oozes
    C.   silica oozes
    D.   all of these
Answer = D (pages 449, 451, 452)

48.   Pelagic sediments _____.
    A.   are fine-grained
    B.   are deposited far from continental margins
    C.   settle very slowly to the seafloor
    D.   all of these
Answer = D (page 449)

49.   Carbonate sediments are deposited at depths down to about _____.
    A.   200 meters
    B.   1 kilometer
    C.   4 kilometers
    D.   10 kilometers
Answer = C (page 451)

50.   The shells of diatoms and radiolaria, which accumulate on the abyssal plain, are composed of _____.
    A.   calcium carbonate
    B.   sodium chloride
    C.   iron sulfide
    D.   silica
Answer = D (page 452)

18

Earthquakes

1. Which of the following describes the build and release of stress during an earth-quake?
    A.   the Modified Mercalli Scale
    B.   the elastic rebound theory
    C.   the principle of superposition
    D.   the seismic gap method
    Answer = B (page 460)

2. Detailed studies of what earthquake allowed researchers to develop the elastic rebound theory?
    A.   the 1906 San Francisco, California earthquake
    B.   the 1964 Anchorage, Alaska earthquake
    C.   the 1755 Lisbon, Portugal earthquake
    D.   the 1985 Mexico City, Mexico earthquake
    Answer = A (page 460)

3. The amount of ground displacement in an earthquake is called the _____.
    A.   epicenter
    B.   dip
    C.   slip
    D.   focus
    Answer = C (page 461)

4.   The geographic point on the Earth's surface directly above the point where slip initiated during an earthquake is called the _____.
   A.   focus
   B.   epicenter
   C.   strike
   D.   dip
Answer = B (page 461)

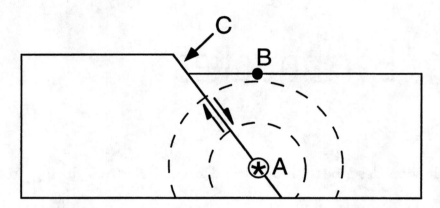

5.   Point A, where slip initiated during the earthquake, is called the _____.
   A.   dip
   B.   epicenter
   C.   focus
   D.   strike
Answer = C (page 461)

6.   Point B is called the earthquake _____.
   A.   dip
   B.   epicenter
   C.   focus
   D.   strike
Answer = B (page 461)

7.   What is the feature labeled "C" in the diagram?
   A.   the earthquake epicenter
   B.   the fault scarp
   C.   the seismic wave
   D.   the earthquake dip
Answer = B (page 461)

8. What type of faulting is illustrated in the diagram?
    A.   normal faulting
    B.   reverse faulting
    C.   strike-slip faulting
    D.   none of the above
    Answer = A (page 469)

9. During a major earthquake, the maximum rupture length along a fault is about
    _____.
    A.   1 kilometer
    B.   10 kilometers
    C.   100 kilometers
    D.   1000 kilometers
    Answer = D (page 462)

Seismographic recording (seismogram) of an earthquake

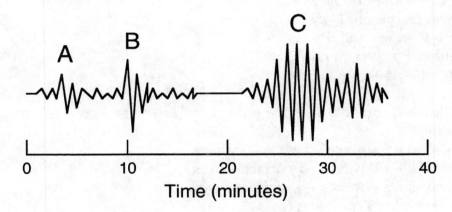

10. What causes the up-and-down wiggles on the seismogram?
     A.   variations in air pressure
     B.   ground vibrations
     C.   tsunami waves
     D.   electromagnetic pulses
     Answer = B (page 463)

11. Which set of waves are probably surface waves?
     A.   set A
     B.   set B
     C.   set C
     D.   sets A, B, and C are all surface waves
     Answer = C (page 464)

12.   How far away from the seismograph station was the earthquake?
      A.   approximately 5 km
      B.   approximately 50 km
      C.   approximately 500 km
      D.   approximately 5000 km
Answer = D (page 466)

13.   Which of the following types of seismic waves arrive at a seismograph first?
      A.   P waves
      B.   S waves
      C.   surface waves
      D.   all of these waves arrive at the same time
Answer = A (page 464)

14.   At a seismograph station, what type of seismic waves arrive before the surface waves?
      A.   P waves, but not S waves
      B.   S waves, but not P waves
      C.   both P waves and S waves
      D.   neither P waves nor S waves
Answer = C (page 464)

15.   Which of the following correctly lists the order in which seismic waves arrive at a seismograph station?
      A.   P waves --> surface waves --> S waves
      B.   S waves --> P waves --> surface waves
      C.   P waves --> S waves --> surface waves
      D.   surface waves --> P waves --> S waves
Answer = C (page 464)

16.   If an earthquake occurred on the San Andreas fault in California, how long would it take for the first seismic wave to arrive at a seismograph station in New York?
      A.   about 15 seconds
      B.   about 10 minutes
      C.   about 5 hours
      D.   about one day
Answer = B (page 464)

17.   How do rock particles move during the passage of a P wave through the rock?
      A.   back and forth parallel to the direction of wave travel
      B.   back and forth at right angles to the direction of wave travel
      C.   in a rolling elliptical motion
      D.   in a rolling circular motion
Answer = A (page 465)

18.   How do rock particles move during the passage of an S wave through the rock?
       A.   back and forth parallel to the direction of wave travel
       B.   back and forth at right angles to the direction of wave travel
       C.   in a rolling elliptical motion
       D.   in a rolling circular motion
       Answer = B (page 465)

19.   How many seismograph stations are needed to locate the epicenter of an earthquake?
       A.   1
       B.   2
       C.   3
       D.   4 or more
       Answer = C (page 466)

20.   The time interval between the arrival of the P and S waves can be used to determine
       _____.
       A.   the location of the earthquake
       B.   the magnitude of the earthquake
       C.   the type of faulting that occurred during the earthquake
       D.   the distance to the earthquake
       Answer = D (page 467)

21.   Who developed the procedure used to measure the size of an earthquake?
       A.   Charles Richter
       B.   James Hutton
       C.   Charles Darwin
       D.   none of these
       Answer = A (page 467)

22.   The energy released during a magnitude 8 earthquake is approximately _____ times
       large than the energy released during a magnitude 6 earthquake.
       A.   2
       B.   10
       C.   100
       D.   1000
       Answer = D (page 467)

23.   The moment magnitude of an earthquake depends on all of the following except
       A.   the area of the fault break
       B.   the rigidity of the rock
       C.   the slip on the fault
       D.   the type of faulting
       Answer = D (page 468)

24.   Which of the following measures an earthquake's intensity based on the observed effects on people and structures?
      A.   the Richter scale
      B.   the Beaufort scale
      C.   the Modified Mercalli scale
      D.   the moment magnitude
    Answer = C (page 468)

25.   What type of information can be gained by examining the first motion (push or pull) of P waves arriving at different seismograph stations?
      A.   the moment magnitude
      B.   the Richter magnitude
      C.   the type of faulting
      D.   the amount of slip
    Answer = C (page 469)

26.   On what type of faults do earthquakes occur?
      A.   normal faults
      B.   reverse faults
      C.   strike-slip faults
      D.   all of these
    Answer = D (page 469)

27.   Shallow earthquakes, less than 20 km deep, are associated with _____.
      A.   convergent plate boundaries
      B.   divergent plate boundaries
      C.   transform plate boundaries
      D.   all of these
    Answer = D (page 470)

28.   Which of the following countries has the least risk of earthquakes?
      A.   Australia
      B.   China
      C.   Japan
      D.   United Sates
    Answer = A (page 470)

29.   Approximately what percentage of earthquakes occur at plate boundaries?
      A.   20%
      B.   50%
      C.   75%
      D.   90%
    Answer = D (page 470)

30.   What type of faulting would be least likely to occur along the mid-Atlantic ridge?
        A.   normal faulting
        B.   reverse faulting
        C.   strike-slip faulting
        D.   all of these types of faulting are likely to occur
Answer = B (page 471)

31.   What type of faulting mechanism is shown by earthquakes that occur along transform faults?
        A.   normal
        B.   reverse
        C.   strike-slip
        D.   none of these
Answer = C (page 471)

Mid-ocean ridge

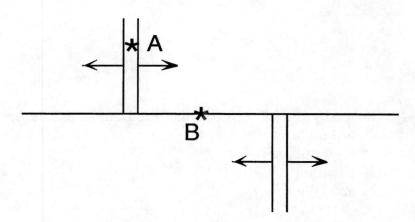

32.   What type of earthquakes would most likely occur at point A?
        A.   shallow-focus earthquakes caused by normal faulting
        B.   shallow-focus earthquakes caused by strike-slip faulting
        C.   shallow-focus earthquakes caused by thrust faulting
        D.   deep-focus earthquakes caused by thrust faulting
Answer = A (page 471)

33.   What type of earthquakes would most likely occur at point B?
        A.   shallow-focus earthquakes caused by normal faulting
        B.   shallow-focus earthquakes caused by strike-slip faulting
        C.   shallow-focus earthquakes caused by thrust faulting
        D.   deep-focus earthquakes caused by thrust faulting
Answer = B (page 471)

34.   Which of the following major earthquakes did not occur at a plate boundary?
      A.   New Madrid, Missouri— 1812
      B.   San Francisco, California—1906
      C.   Anchorage, Alaska—1964
      D.   Loma Prieta, California—1989
Answer = A (page 471)

35.   How often do magnitude 8 earthquakes occur?
      A.   about 5 to 10 times per year
      B.   about once a year
      C.   about once every 5 to 10 years
      D.   about once every 50 to 100 years
Answer = C (page 472)

36.   Approximately how many people lost their lives during the 1989 Loma Prieta (San Francisco) earthquake?
      A.   6
      B.   60
      C.   600
      D.   6,000
Answer = B (page 472)

37   How much damage was caused by the 1994 Northridge earthquake in Los Angeles?
      A.   approximately $20 million
      B.   approximately $200 million
      C.   approximately $2 billion
      D.   approximately $20 billion
Answer = D (page 472)

38.   Approximately how many earthquakes with Richter magnitudes between 6 and 7 occur each year?
      A.   1
      B.   100
      C.   10,000
      D.   1,000,000
Answer = B (page 472)

39.   Which of the following earthquakes was the smallest?
      A.   the 1906 San Francisco earthquake
      C.   the 1923 Tokyo earthquake
      B.   the 1964 Alaska earthquake
      C.   the 1994 Northridge earthquake
Answer = D (page 473)

40.   Which of the following can be triggered by an earthquake?
      A.   a tsunami
      B.   intense ground shaking
      C.   a landslide
      D.   all of these
Answer = D (pages 473–474)

41.   Tsunamis can be generated by _____.
      A.   undersea earthquakes
      B.   undersea landslides
      C.   the eruption of an oceanic volcano
      D.   all of these
Answer = D (page 474)

42.   Which of the following types of waves are slowest?
      A.   P waves
      B.   S waves
      C.   surface waves
      D.   tsunamis
Answer = D (page 474)

43.   How fast do tsunamis travel across the ocean?
      A.   up to 5 kilometers per hour
      B.   up to 60 kilometers per hour
      C.   up to 800 kilometers per hour
      D.   up to 10,000 kilometers per hour
Answer = C (page 475)

44.   Which of the following best describes the expected level of earthquake-shaking hazards in Tempe, Arizona [enter your city]?
      A.   low (< 2% g)
      B.   moderate (2–8% g)
      C.   high (8–24% g)
      D.   extreme (> 24% g)
Answer = B for Tempe, Arizona (page 477)

45.   Which of the following statements about earthquakes is false?
      A.   Most earthquakes occur at plate boundaries.
      B.   The time and location of most major earthquakes can be predicted several days in advance.
      C.   Earthquakes can be caused by normal, reverse, and strike-slip faulting.
      D.   P waves travel faster than both S waves and surface waves.
Answer = B (pages 478–479)

46.   Which of the following observations may indicate a forthcoming destructive earthquake?
        A.   an increase in the frequency of smaller earthquakes in the region
        B.   rapid tilting of the ground
        C.   rapid changes in water levels in wells
        D.   all of these
Answer = D (pages 479–480)

47.   Which of the following statements best describes the state of earthquake prediction?
        A.   Scientists can accurately predict the time and location of almost all earthquakes.
        B.   Scientists can accurately predict the time and location of about 50% of all earthquakes.
        C.   Scientists can accurately predict when an earthquake will occur, but not where.
        D.   Scientists can characterize the seismic risk of an area, but can not yet accurately predict most earthquakes.
Answer = D (page 481)

48.   In southern California, great earthquakes occur on the San Andreas approximately every _____ years.
        A.   1–2
        B.   10–15
        C.   100–150
        D.   1000
Answer = C (page 481)

49.   The last great earthquake on the San Andreas fault in southern California occurred in _____.
        A.   1857
        B.   1906
        C.   1989
        D.   1994
Answer = A (page 481)

50.   On what type of fault did the 1994 Northridge earthquake in southern California occur?
        A.   a normal fault
        B.   a strike-slip fault
        C.   a thrust fault
        D.   scientists do not know because the fault did not rupture the surface
Answer = C (page 481)

19

Exploring Earth's Interior

1.   What is the approximate distance from the surface to the center of the Earth?
      A.   1,000 miles
      B.   4,000 miles
      C.   10,000 miles
      D.   40,000 miles
      Answer = B (page 485)

2.   When seismic waves pass through the boundary between two different materials, _____.
      A.   the waves refract (bend)
      B.   the velocity of the waves changes
      C.   both A and B
      D.   neither A nor B
      Answer = C (page 486)

Cross section of the Earth

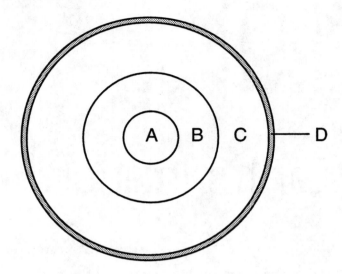

3.   Which region in the Earth is molten?
        A.   region A
        B.   region B
        C.   region C
        D.   region D
    Answer = B (page 487)

4.   Which region in the Earth consists primarily of solid iron?
        A.   region A
        B.   region B
        C.   region C
        D.   region D
    Answer = A (page 493)

5.   In which region of the Earth is conduction the dominant heat transfer mechanism?
        A.   region A
        B.   region B
        C.   region C
        D.   region D
    Answer = D (page 496)

6.   How long does it take a P-wave to travel through the Earth?
        A.   approximately 20 seconds
        B.   approximately 1 minute
        C.   approximately 5 minutes
        D.   approximately 20 minutes
    Answer = D (page 487)

7.   Which of the following statements is false?
       A.   The P-wave shadow zone is larger than the S-wave shadow zone.
       B.   Seismic waves follow curved paths through the interior of the Earth.
       C.   P waves travel more slowly in the outer core than in the lower mantle.
       D.   Liquids do not transmit S waves.
      Answer = A (page 487)

8.   The S-wave shadow zone is caused by _____.
       A.   the crust-mantle boundary
       B.   the outer core
       C.   the lower mantle
       D.   the inner core
      Answer = B (page 487)

9.   The S-wave shadow zone extends from _____ to 180° angular distance from the earth-
      quake focus.
       A.   45°
       B.   75°
       C.   105°
       D.   145°
      Answer = C (page 487)

Cross section through the Earth

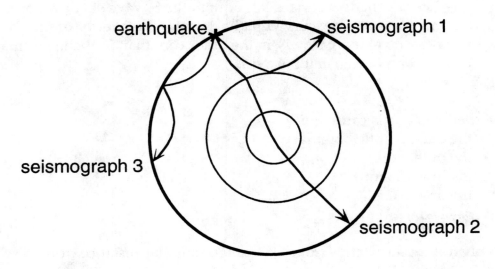

10. Which of the following best describes the compressional seismic wave arriving at seismograph 1?
    A.   P wave
    B.   PP wave
    C.   PcP wave
    D.   PKP wave
Answer = C (pages 487-488)

11. Which of the following statements regarding the seismic wave arriving at seismograph 2 is false?
    A.   The wave is referred to as a PKIKP wave.
    B.   The wave passes through the inner core.
    C.   The wave is refracted at the major boundaries in the Earth.
    D.   The wave could be either a P wave or an S wave.
Answer = D (page 487)

12. Which of the following best describes the shear wave arriving at seismograph 3?
    A.   PcP
    B.   PP
    C.   ScS
    D.   SS
Answer = D (page 487)

13.   S waves that reflect off the Earth's surface back into the Earth are called _____.
      A.   SP
      B.   SS
      C.   ScS
      D.   Sp
Answer = B (page 487)

14.   What type of P-wave reflects off the core back to the Earth's surface?
      A.   PcP
      B.   PP
      C.   PKP
      D.   none of these
Answer = A (page 488)

15.   The boundary between the mantle and the core lies at a depth of approximately _____.
      A.   300 kilometers
      B.   1000 kilometers
      C.   3000 kilometers
      D.   10,000 kilometers
Answer = C (page 488)

16.   Continental crust beneath mountains can be up to ___ kilometers thick.
      A.   5
      B.   35
      C.   65
      D.   100
Answer = C (page 488)

17.   Where do P-waves travel fastest?
      A.   in the upper mantle
      B.   in the lower mantle
      C.   in the outer core
      D.   in the inner core
Answer = B (page 489)

18.   What region of the Earth takes up the greatest volume?
      A.   the crust
      B.   the outer core
      C.   the inner core
      D.   the mantle
Answer = D (page 489)

19.   The continental crust consists mostly of _____.
        A.   granitic rocks
        B.   gabbroic rocks
        C.   ultramafic rocks
        D.   sedimentary rocks
        Answer = A (page 489)

20.   How fast do P-waves travel through peridotite?
        A.   4 kilometers per second
        B.   6 kilometers per second
        C.   8 kilometers per second
        D.   10 kilometers per second
        Answer = C (page 489)

21.   Which of the following statements about the Moho (Mohovoricic discontinuity) is false?
        A.   Seismic waves speed up as they pass down through the Moho.
        B.   The Moho separates denser rocks below from less dense rocks above.
        C.   The Moho separates the crust from the mantle.
        D.   The Moho marks the top of a partially molten layer.
        Answer = D (page 489)

22.   The boundary between the Earth's crust and mantle was first discovered by _____.
        A.   analyzing seismic waves
        B.   deep continental drilling
        C.   detailed geologic mapping
        D.   paleomagnetic studies
        Answer = A (page 489)

23.   How fast do P-waves travel through granite?
        A.   4 kilometers per second
        B.   6 kilometers per second
        C.   8 kilometers per second
        D.   10 kilometers per second
        Answer = B (page 489)

24.   Which of the following statements is false?
        A.   The crust-mantle boundary is called the Mohorovicic discontinuity.
        B.   The oceanic crust consists of basalt and gabbro.
        C.   The crust is less dense than the mantle.
        D.   P-waves travel faster in the crust than in the mantle.
        Answer = D (page 489)

25.   Where in the Earth are P-wave velocities the highest?
        A.   the upper mantle
        B.   the lower mantle
        C.   the outer core
        D.   the inner core
        Answer = B (page 489)

26.   After melting of a continental ice cap, the surface of the continent will tend to _____.
        A.   rise
        B.   sink
        C.   rise or sink depending on the thickness of the ice cap
        D.   remain the same
        Answer = A (page 490)

27.   Which of the following regions in the Earth consists primarily of olivine and pyroxene?
        A.   the upper mantle
        B.   the lower mantle
        C.   the outer core
        D.   the inner core
        Answer = A (page 491)

28.   Which of the following statements is true?
        A.   The lithosphere contains the crust.
        B.   The crust contains the lithosphere.
        C.   The lithosphere and crust are different terms for the same part of the Earth.
        D.   The lithosphere and crust are separate parts of the Earth.
        Answer = A (page 491)

29.   The lithosphere is approximately ___ kilometers thick.
        A.   25
        B.   100
        C.   250
        D.   1000
        Answer = B (page 491)

30.   Which of the following statements about the asthenosphere is false?
        A.   The asthenosphere lies beneath the lithosphere.
        B.   The asthenosphere is stronger than the lithosphere.
        C.   The asthenosphere rises close to the surface beneath mid-ocean ridges.
        D.   The asthenosphere is partially molten.
        Answer = B (page 491)

31.   Who discovered that the inner core was solid?
    A.   Inge Lehman
    B.   Mohorovicíc
    C.   Charles Richter
    D.   Alfred Wegener
Answer = A (page 492)

32.   The sharp increases in the velocity of S-waves at 400 and 670 kilometers depth in the mantle are probably caused by _____.
    A.   changes to more compact mineral structures
    B.   changes in the composition of the mantle
    C.   changes in the temperature of the mantle
    D.   changes in the pressure of the mantle
Answer = A (page 492)

33.   What element makes up most of the Earth's core?
    A.   silicon
    B.   oxygen
    C.   iron
    D.   magnesium
Answer = C (page 492)

34.   The boundary between the inner core and the outer core lies at a depth of _____.
    A.   700 kilometers
    B.   2900 kilometers
    C.   5100 kilometers
    D.   6400 kilometers
Answer = C (page 492)

35.   How thick a continental root would be produced by a 3-km-thick continental ice sheet?
    A.   about 1 km
    B.   about 3 km
    C.   about 10 km
    D.   about 30 km
Answer = A (page 493)

36.   Which of the following statements best describes the nature of the core-mantle boundary?
        A.   The core-mantle boundary is smooth.
        B.   The core-mantle boundary is rough with a topography of about 5 km.
        C.   The core-mantle boundary is rough with a topography of about 100 km.
        D.   The core-mantle boundary is rough with a topography of about 400 km.
Answer = B (page 494)

37.   Which of the following statements regarding the inner core is true?
        A.   P waves do not travel through the inner core.
        B.   P waves travel faster through the inner core in a north-south direction than in an east-west direction.
        C.   P waves travel faster through the inner core in an east-west direction than in a north-south direction.
        D.   P waves travel faster through the inner core at the same speed regar less of direction.
Answer = B (page 494)

38.   The mechanical transfer of heat by vibration of atoms and molecules is called _____.
        A.   radiation
        B.   conduction
        C.   magnetism
        D.   convection
Answer = B (page 495)

39.   If the Earth cooled only by conduction, heat from depths greater than _____ kilometers would not yet have reached the surface.
        A.   20
        B.   100
        C.   400
        D.   2000
Answer = C (page 496)

40.   What drives plate tectonics?
        A.   thermal convection
        B.   thermal conduction
        C.   solar energy
        D.   erosion
Answer = A (page 496)

41.   In a deep mine, temperatures increase at the rate of _____.
        A.   approximately 2.5°C per kilometer
        B.   approximately 25°C per kilometer
        C.   approximately 250°C per kilometer
        D.   approximately 2500°C per kilometer
Answer = B (page 497)

42.   Who first proposed that the Earth acted as a large magnet whose field forces the nee-
        dle of a magnetic compass to align north-south?
        A.   Charles Lyell
        B.   Charles Darwin
        C.   James Hutton
        D.   William Gilbert
Answer = D (page 498)

43.   What is the name for a curve that shows how temperature changes with depth in the
        Earth?
        A.   a geotherm
        B.   a travel-time curve
        C.   a melting curve
        D.   an isograd
Answer = A (page 498)

44.   Where is the Earth's magnetic field generated?
        A.   in the crust
        B.   in the mantle
        C.   in the outer core
        D.   in the inner core
Answer = C (page 498)

45.   Above what temperature do materials lose their permanent magnetism?
        A.   50°C
        B.   100°C
        C.   250°C
        D.   500°C
Answer = D (page 498)

46.   Permanent magnetism acquired by minerals in igneous rocks during crystallization
        is called _____ magnetization.
        A.   depositional remanent
        B.   plutonoremanent
        C.   silicate
        D.   thermoremanent
Answer = D (page 499)

47. Which of the following statements is true?
   A. The Earth's magnetic poles are aligned with the Earth's rotation axis.
   B. The Earth's magnetic poles are inclined approximately 10° from the Earth's rotation axis.
   C. The Earth's magnetic poles are inclined approximately 45° from the Earth's rotation axis.
   D. The Earth's magnetic poles are perpendicular to the Earth's rotation axis.
   Answer = B (page 499)

48. Which of the following rock types would be most likely to record the magnetic field at the time the rock formed?
   A. an alluvial conglomerate
   B. a basaltic lava flow
   C. an evaporite deposit of halite
   D. a metamorphic schist
   Answer = B (page 500)

49. The Earth has had a magnetic field for _____.
   A. the last 5 million years
   B. the last 100 million years
   C. the last 1 billion years
   D. the last 3.5 billion years
   Answer = D (page 500)

50. The Earth's magnetic field reverses itself roughly every _____.
   A. 50 years
   B. 5,000 years
   C. 500,000 years
   D. 50 million years
   Answer = C (page 501)

20

Plate Tectonics:
The Unifying Theory

1. What is the name of the Mesozoic supercontinent that consisted of all of the present continents?
    A.   Eurasia
    B.   Laurasia
    C.   Pangaea
    D.   Gondwanaland
    Answer = C (page 506)

2. When did the supercontinent Pangaea begin to break apart?
    A.   about 10,000 years ago
    B.   about 10 million years ago
    C.   about 200 million years ago
    D.   about 570 million years ago
    Answer = C (page 506)

3. What two scientists proposed the theory of seafloor spreading in the early 1960s?
    A.   Charles Darwin and James Hutton
    B.   Harry Hess and Robert Dietz
    C.   F. J. Vine and D. H. Mathews
    D.   Alfred Wegener and Arthur Holmes
    Answer = B (page 507)

4.   What age are the fossils of the reptile Mesosaurus found in Africa and South America that suggest the two continents were once together?
        A.   early Cenozoic
        B.   late Mesozoic
        C.   early Mesozoic
        D.   late Paleozoic
    Answer = D (page 507)

5.   The theory of plate tectonics was widely accepted by _____.
        A.   the end of the 19th century
        B.   about 1930
        C.   about 1950
        D.   about 1970
    Answer = D (page 508)

6.   How old are the oldest rocks on the ocean floor?
        A.   about 20 million years old
        B.   about 175 million years old
        C.   about 570 million years old
        D.   about 4,000 million years old
    Answer = B (page 508)

7.   Continental rocks provide a fragmentary record of _____ of Earth's history.
        A.   approximately 5%
        B.   approximately 20%
        C.   approximately 50%
        D.   approximately 90%
    Answer = D (page 508)

8.   Approximately how many major lithospheric plates are there?
        A.   3
        B.   10
        C.   30
        D.   100
    Answer = B (page 508)

9.   New oceanic lithosphere forms at _____.
        A.   convergent plate boundaries
        B.   divergent plate boundaries
        C.   transform plate boundaries
        D.   all of these
    Answer = B (page 509)

10. Which of the following is not a divergent plate boundary?
     A. the Great Rift Valley of East Africa
     B. the East Pacific Rise
     C. the San Andreas fault
     D. the Mid-Atlantic Ridge
   Answer = C (page 509)

11. Partial melting of the mantle takes place at _____.
     A. divergent plate boundaries
     B. ocean-ocean convergent plate boundaries
     C. ocean-continent convergent plate boundaries
     D. all of these
   Answer = D (pages 509-511)

12. At what type of plate boundary do shallow-focus earthquakes occur?
     A. convergent
     B. divergent
     C. transform
     D. all of these
   Answer = D (page 509-512)

13. The Hawaiian Islands formed at a _____.
     A. convergent plate boundary
     B. divergent plate boundary
     C. transform plate boundary
     D. none of these
   Answer = D (page 510)

14. Which of the following are not associated with convergent plate margins?
     A. deep-focus earthquakes
     B. rift valleys
     C. island arcs
     D. deep-sea trenches
   Answer = B (pages 510-511)

15. What is the name of the plate that is subducting beneath western South America?
     A. the Pacific plate
     B. the South American plate
     C. the Nazca plate
     D. the Atlantic plate
   Answer = C (page 511)

16.   Deep-sea trenches are associated with what type of plate boundary?
        A.   convergent
        B.   divergent
        C.   transform
        D.   all of these
        Answer = A (page 511)

17.   Which of the following is an example of a transform plate boundary?
        A.   the East Pacific Rise
        B.   the San Andreas Fault
        C.   the Mid-Atlantic Ridge
        D.   the East African Rift
        Answer = B (page 512)

18.   The North American plate is bounded by _____ plate boundaries.
        A.   convergent
        B.   divergent
        C.   transform
        D.   convergent, divergent, and transform
        Answer = D (page 512)

19.   Which of the following increases with distance from a mid-ocean ridge?
        A.   the age of the oceanic lithosphere
        B.   the depth to the sea floor
        C.   the thickness of the lithosphere
        D.   all of the above
        Answer = D (page 513)

20.   Plate tectonic rates were first calculated in the early 1960s by _____.
        A.   dating rocks drilled from the seafloor
        B.   using seafloor magnetic anomalies
        C.   global-positioning satellite technology
        D.   fossil evidence
        Answer = B (page 514)

The cross section depicts magnetized oceanic crust at a spreading center. The "+" symbol indicates normal magnetic bands; the "−" symbol indicates reversed magnetic bands.

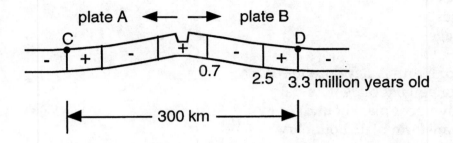

21.   How many reversals of the Earth's magnetic field are depicted in the diagram?
         A.   3
         B.   4
         C.   6
         D.   7
Answer = A (page 515)

22.   How fast are points C and D spreading apart from each other?
         A.   about 2 centimeters/year
         B.   about 5 centimeters/year
         C.   about 10 centimeters/year
         D.   about 20 centimeters/year
Answer = C (page 515)

23.   "Normal" magnetized crust at the spreading center formed during the _____ epoch.
         A.   Brunhes
         B.   Gauss
         C.   Gilbert
         D.   Matuyama
Answer = A (page 515)

24.   When did deep-sea drilling of the seafloor begin?
         A.   around 1900
         B.   in the 1950s
         C.   in the late 1960s
         D.   in the early 1990s
Answer = C (page 516)

25.   Lines on the seafloor that connect rocks of the same age are called _____.
          A.   isograds
          B.   isotopes
          C.   isochrons
          D.   isostasy
       Answer = C (page 517)

26.   The east coast of North America represents a _____.
          A.   convergent plate boundary
          B.   divergent plate boundary
          C.   transform plate boundary
          D.   none of these
       Answer = D (page 518)

27.   How fast is North America separating from Eurasia?
          A.   approximately 2 centimeters per year
          B.   approximately 10 centimeters per year
          C.   approximately 50 centimeters per year
          D.   approximately 2 meters per year
       Answer = A (page 518)

28.   What are ophiolite suites?
          A.   fragments of oceanic lithosphere emplaced on a continent
          B.   groups of seafloor magnetic anomalies
          C.   wedge-shaped packages of sediments that form at passive margins
          D.   micro-continents that have traveled a long distance
       Answer = A (page 520)

Idealized cross section through an ophiolite suite

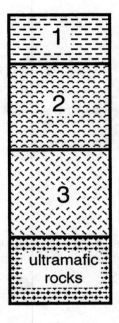

29. Which of the following rock types would you least expect to find in layer 1?
    A. chert
    B. limestone
    C. sandstone
    D. shale
    Answer = C (page 521)

30. What type of rocks make up layer 2?
    A. basalt
    B. gabbro
    C. granite
    D. peridotite
    Answer = A (page 521)

31. Where is the Moho in the diagram?
    A. between layers 1 and 2
    B. between layers 2 and 3
    C. between layer 3 and the ultramafic rocks
    D. at the base of the ultramafic rocks
    Answer = C (page 521)

32.   Which two layers have the same chemical composition?
        A.   layers 1 and 2
        B.   layers 1 and 3
        C.   layers 2 and 3
        D.   layer 3 and the ultramafic rocks
      Answer = C (page 521)

33.   How wide are the crystal mush zones (magma chambers) beneath mid-ocean ridges?
        A.   approximately 1 kilometer
        B.   approximately 10 kilometers
        C.   approximately 100 kilometers
        D.   approximately 1000 kilometers
      Answer = B (page 521)

34.   Continental shelf deposits may accumulate to thicknesses of approximately
      _____.
        A.   100 meters
        B.   1 kilometer
        C.   10 kilometers
        D.   100 kilometers
      Answer = C (page 522)

35.   Sedimentary rocks that are laid down in a slowly subsiding basin along a receding
      continental margin are called _____.
        A.   continental shelf deposits
        B.   ophiolite suites
        C.   melanges
        D.   fluvial deposits
      Answer = A (page 522)

36.   Which of the following is not a chain of volcanic islands associated with ocean-ocean
      convergence?
        A.   the Aleutian Islands
        B.   the Hawaiian Islands
        C.   the Mariana Islands
        D.   the Philippine Islands
      Answer = B (page 524)

37.   Volcanic island arcs are associated with _____.
        A.   transform plate boundaries
        B.   divergent plate boundaries
        C.   ocean-ocean convergent plate boundaries
        D.   ocean-continent convergent plate boundaries
      Answer = C (page 524)

38.   What is the difference between a mélange belt and a magmatism belt left behind by
        former ocean-continent convergence?
      A.   the type of metamorphism
      B.   the rock types
      C.   the structural features
      D.   all of these
        Answer = D (page 525)

Cross section of an ocean-continent plate boundary

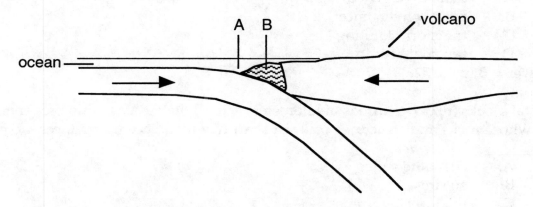

39.   What is the topographic feature labeled A?
      A.   an oceanic basin
      B.   an oceanic channel
      C.   an oceanic rift
      D.   an oceanic trench
        Answer = D (page 525)

40.   Feature B, called a _____, consists of chaotically mixed and deformed rocks.
      A.   forearc basin
      B.   mélange
      C.   suture
      D.   turbidite
        Answer = B (page 525)

41.   What type of metamorphism occurs in region B?
      A.   high temperature, low pressure
      B.   low temperature, high pressure
      C.   high temperature, high pressure
      D.   both A and B
        Answer = B (page 525)

42.   Which of the following locations could be represented by the diagram?
          A.   the east coast of Africa
          B.   the east coast of North America
          C.   the west coast of South America
          D.   the west coast of Europe
        Answer = C (page 513)

43.   In California, the Franciscan mélange formation and the Sierra Nevada batholith
        mark the location of former _____.
          A.   plate divergence
          B.   plate convergence
          C.   transform faulting
          D.   hot spot activity
        Answer = B (page 526)

44.   Crustal blocks (up to hundreds of kilometers across) that occur within orogenic belts
        and whose rocks and structures contrast sharply with adjacent provinces are called

        _____.
          A.   ophiolite suites
          B.   sutures
          C.   microplate terranes
          D.   island arcs
        Answer = C (page 526)

45.   When did India begin to collide with Asia to form the Himalayas?
          A.   about 50 million years ago
          B.   about 200 million years ago
          C.   about 500 million years ago
          D.   about 2,000 million years ago
        Answer = A (page 526)

46.   Over geologic time, the Earth's continents have grown at an average rate of approx-
        imately _____ cubic kilometers per year.
          A.   2
          B.   200
          C.   20,000
          D.   2,000,000
        Answer = A (page 528)

47.   Which of the following mountains did not form as a result of collision between two continents?

      A.   Appalachians
      B.   Urals
      C.   Andes
      D.   Himalayas

Answer = C (page 529)

48.   What is the name of the ancient ocean that surrounded Pangaea?

      A.   Panthalassa
      B.   Tethys
      C.   Atlantic
      D.   Laurasia

Answer = A (page 530)

49.   During the breakup of Pangaea, which of these events occurred last?

      A.   the separation of Laurasia from Gondwana
      B.   the collision of India with Asia
      C.   the separation of Africa from South America
      D.   the separation of North America from Eurasia

Answer = B (page 532)

50.   How much of the present-day sea floor was formed during Tertiary time (the past 65 million years)?

      A.   about 25%
      B.   about 50%
      C.   about 75%
      D.   about 100%

Answer = B (page 532)

51.   Over the past 130 million years, how much sea floor has been subducted beneath western North America?

      A.   about 30 kilometers
      B.   about 150 kilometers
      C.   about 900 kilometers
      D.   about 7000 kilometers

Answer = D (page 532)

21

Deformation of the
Continental Crust

1.   Which of the following mountain chains was not the result of a continental collision
      during the assembly of the supercontinent Pangaea?
              A.   Appalachians
              B.   Caledonides
              C.   Himalayas
              D.   Urals
      Answer = C (page 538)

2.   Oceanic crust provides a record of about ___ of Earth's history.
              A.   5%
              B.   25%
              C.   75%
              D.   95%
      Answer = A (page 539)

3.   Approximately how old are the oldest continental rocks?
              A.   50 million years old
              B.   200 million years old
              C.   1.5 billion years old
              D.   4 billion years old
      Answer = D (page 539)

4.   What is the general term for periods of mountain-building that include folding, fault-
       ing, magmatism, and metamorphism?
              A.   rejuvenation
              B.   convergence
              C.   orogeny
              D.   sedimentation
       Answer = C (page 539)

5.   The oldest rocks tend to be found _____.
              A.   in the interiors of ocean basins
              B.   along the margins of ocean basins
              C.   in the interior of continents
              D.   along the margins of continents
       Answer = C (page 539)

6.   What is the name of the mountain chain that runs along the western margin of South
       America?
              A.   Appalachians
              B.   Andes
              C.   Cordillera
              D.   Himalayas
       Answer = B (page 540)

Map of North America

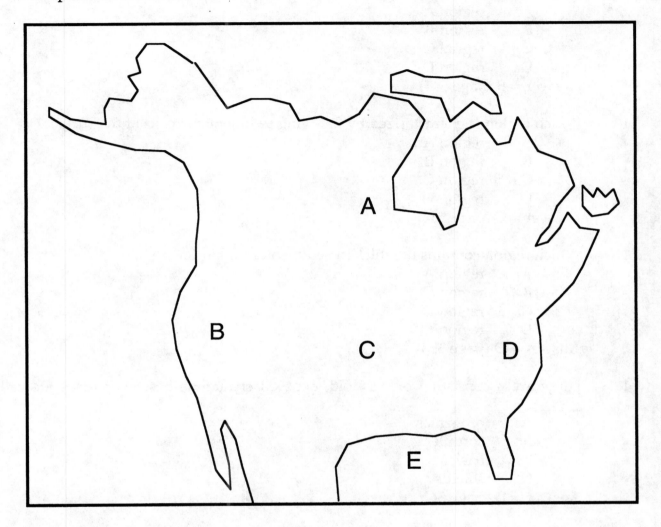

7.   Which region contains the oldest rocks in North America?
         A.   region A
         B.   region B
         C.   region C
         D.   region D
      Answer = A (page 541)

8.   Which region consists of folded and faulted rocks that formed during the Paleozoic era?
         A.   region A
         B.   region B
         C.   region C
         D.   region D
      Answer = D (page 541)

9.   Which region represents the Cordilleran orogenic belt?
           A.   region A
           B.   region B
           C.   region C
           D.   region D
       Answer = B (page 541)

10.   Which region is characterized by flat-lying sedimentary rocks at the surface?
           A.   region A
           B.   region B
           C.   region C
           D.   region D
       Answer = C (page 542)

11.   Which region contains the thickest sedimentary deposits?
           A.   region A
           B.   region C
           C.   region D
           D.   region E
       Answer = D (page 553)

12.   Large areas consisting of very old, exposed crystalline basement rocks are called
       _____.
           A.   orogenies
           B.   cordillera
           C.   basins
           D.   shields
       Answer = D (page 541)

13.   Where are the oldest rocks in North America located?
           A.   in the Canadian Shield
           B.   in the Rocky Mountains
           C.   in the Appalachian Mountains
           D.   in the Basin and Range province
       Answer = A (page 541)

14.   What type of geologic contact separates the rocks of the Canadian Shield from the
       rocks of the Great Plains? (Hint: consider the type and age of the rocks found in
       each area.)
           A.   a depositional unconformity
           B.   a glacial moraine
           C.   a thrust fault
           D.   an intrusive contact
       Answer = A (page 542)

15.　How thick is the sequence of sedimentary rocks that underlies the Great Plains?
　　A.　about 400 meters thick
　　B.　about 2 kilometers thick
　　C.　about 10 kilometers thick
　　D.　about 50 kilometers thick
Answer = B (page 542)

16.　The rock assemblages in the Interior Platform of North America indicate deposition
_____.
　　A.　in extensive shallow seas
　　B.　on alluvial plains
　　C.　in lakes and swamps
　　D.　*all* of these
Answer = D (page 542)

Cross section of a continental collision

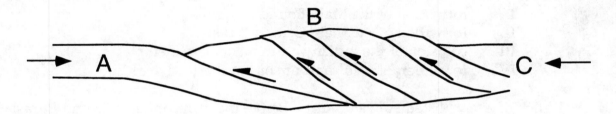

17.　What type of faults are depicted in the cross section?
　　A.　normal faults
　　B.　strike-slip faults
　　C.　thrust faults
　　D.　cannot tell from the information given
Answer = C (page 542)

18.　If B represents the Himalayas, then _____.
　　A.　plate A is Asia and plate C is Europe
　　B.　plate A is India and plate C is Asia
　　C.　plate A is Asia and plate C is India
　　D.　plate A is Europe and plate C is Asia
Answer = B (page 543)

19.   If the cross section depicts the southern Appalachian Mountains approximately 250 million years ago, then _____.
        A.   plate A is Africa and plate C is Europe
        B.   plate A is Africa and plate C is North America
        C.   plate A is North America and plate C is Europe
        D.   plate A is North America and plate C is Africa
Answer = D (page 547)

20.   The Himalayas began to form approximately _____ when the Indian subcontinent began to collide with Tibet.
        A.   5 million years ago
        B.   50 million years ago
        C.   500 million years ago
        D.   5 billion years ago
Answer = B (page 543)

Development of the Himalayas:
        I.   formation of the Main Boundary Thrust
        II.   formation of the Main Central thrust
        III.   collision of the Indian and Eurasian continents
        IV.   formation of a volcanic arc on southern edge of Eurasia

21.   Which of events listed above occurred before the collision of India and Eurasia?
        A.   event I
        B.   event II
        C.   event IV
        D.   events I, II, and IV
Answer = C (page 543)

22.   Which of the following correctly lists the sequence of events that led to the development of the Himalayas?
        A.   I —> II —> III —> IV
        B.   IV —> III —> II —> I
        C.   III —> II —> I —> IV
        D.   II —> I —> IV —> III
Answer = B (page 543)

23.   The Valley and Ridge province of the Appalachian Mountains consists of Paleozoic sedimentary rocks that have been thrust to the _____.
        A.   northeast
        B.   northwest
        C.   southeast
        D.   southwest
Answer = B (page 544)

24.   Which of the following physiographic regions is not part of the Appalachian
       Mountains?
              A.   the Valley and Ridge province
              B.   the Piedmont
              C.   the Blue Ridge province
              D.   the Basin and Range province
       Answer = D (pages 544–546)

25.   Most of the Appalachian Mountains formed during _____ time.
              A.   Cenozoic
              B.   Mesozoic
              C.   Paleozoic
              D.   Precambrian
       Answer = C (pages 544–546)

26.   Which of the following physiographic regions in the Appalachian Mountains is the
       most intensely deformed?
              A.   the Appalachian plateaus
              B.   the Blue Ridge province
              C.   the Piedmont
              D.   the Valley and Ridge Province
       Answer = C (page 545)

27.   In the Appalachian Mountains, a second orogenic period occurred in _____ time
       when an island arc collided with North America.
              A.   Ordovician
              B.   Jurassic
              C.   Carboniferous-Permian
              D.   Devonian
       Answer = D (page 546)

28.   When did Africa collide with North America?
              A.   during Ordovician time
              B.   during Jurassic time
              C.   during Carboniferous-Permian time
              D.   during Devonian time
       Answer = C (page 546)

29.   The Appalachian Mountains consist of _____.
              A.   deformed continental shelf sediments
              B.   an accreted island arc
              C.   an accreted continental fragment
              D.   all of these
       Answer = D (page 546)

30.   The thrust sheets of the Blue Ridge and Piedmont provinces moved approximately _____ over the continental shelf sediments of ancient North America.
         A.   250 meters
         B.   2.5 kilometers
         C.   25 kilometers
         D.   250 kilometers
     Answer = D (page 546)

31.   When did the modern North Atlantic Ocean begin to open?
         A.   approximately 100 million years ago
         B.   approximately 200 million years ago
         C.   approximately 300 million years ago
         D.   approximately 400 million years ago
     Answer = B (page 546)

32.   Which of the following mountain belts formed as a result of a collision between two continents?
         A.   Appalachians
         B.   Himalayas
         C.   Urals
         D.   all of these
     Answer = D (page 546)

33.   Which of the following mountain belts is a mirror image of the Appalachians?
         A.   Alps
         B.   Caledonides
         C.   Himalayas
         D.   Mauritanides
     Answer = D (page 546)

34.   The North America Cordillera formed by the collision of North America and _____.
         A.   Asia
         B.   South America
         C.   Africa
         D.   numerous continental and oceanic fragments
     Answer = D (page 548)

35.   Which of the following regions contains numerous fault-block mountain ranges?
         A.   the Appalachians
         B.   the Basin and Range province
         C.   the Canadian Shield
         D.   the Great Plains
     Answer = B (page 548)

36.   The Front Range of the Rocky Mountains in Colorado is an example of a _____.
        A.   fault block mountain
        B.   folded mountain
        C.   upwarped mountain
        D.   volcanic mountain
        Answer = C (page 549)

37.   Which of the following did not form as a result of the collision between Asia and India?
        A.   the Himalayas
        B.   the Tibetan Plateau
        C.   the Altyn Tagh strike-slip fault
        D.   the Ural Mountains
        Answer = D (page 550)

38.   The main thrust sheets in the Himalayas were transported _____.
        A.   northward
        B.   southward
        C.   eastward
        D.   westward
        Answer = B (page 550)

39.   The North American Cordillera formed over a period of _____.
        A.   10 million years
        B.   50 million years
        C.   250 million years
        D.   1 billion years
        Answer = D (page 550)

40.   The topography of the Basin and Range province in western North America is caused by _____.
        A.   normal faulting
        B.   thrust faulting
        C.   upwarping
        D.   strike-slip faulting
        Answer = A (page 552)

41.   Which of the following events occurred most recently?
        A.   development of the Basin and Range province
        B.   initial rifting of the Atlantic Ocean
        C.   magmatism in the Canadian Shield
        D.   collision of North America and Africa
        Answer = A (page 552)

42.   Which of the following is an example of a fault-block mountain range?
         A.   the Adirondacks of New York
         B.   the Cascade Mountains of western Oregon and Washington
         C.   the Teton Range of Wyoming
         D.   all of these
      Answer = C (page 552)

43.   Which of the following rivers runs along a rift valley?
         A.   the Mississippi River
         B.   the Connecticut River
         C.   the Colorado River
         D.   the Ohio River
      Answer = B (page 552)

44.   How thick are the sedimentary rocks in the Gulf of Mexico?
         A.   up to 100 meters thick
         B.   up to 1 kilometer thick
         C.   up to 10 kilometers thick
         D.   up to 100 kilometers thick
      Answer = C (page 553)

45.   Gradual vertical movements of the crust without significant deformation are called
      _____.
         A.   fault blocks
         B.   orogeny
         C.   cratons
         D.   epeirogeny
      Answer = D (page 553)

46.   In which of the following regions would you expect to find the thickest sequence of
      sedimentary rocks?
         A.   on continental platforms
         B.   along mid-ocean ridges
         C.   on continental shelves
         D.   in the middle of the ocean basin, away from a ridge
      Answer = C (page 553)

47.   Which of the following is an example of an epeirogenic sedimentary basin?
         A.   the Connecticut River valley
         B.   the Michigan Basin
         C.   the Atlantic coastal plain
         D.   none of these
      Answer = B (page 554)

48. Which of the following mountains formed by general upward movement without major faulting?
    A. the Teton Range of Wyoming
    B. the Black Hills of South Dakota
    C. the Sierra Nevada of California
    D. the Cascade Mountains of Oregon and Washington
    Answer = B (page 554)

49. Which part of the United States is currently being uplifted at a rate of over 10 millimeters per year?
    A. New England
    B. Gulf Coast of Texas and Louisiana
    C. northern Great Plains
    D. California
    Answer = C (page 556)

50. Which of the following statements regarding Arizona [your state] is true?
    A. Arizona is subsiding at a rate of 5–10 mm/yr
    B. Arizona is subsiding at a rate of 1–5 mm/yr
    C. Arizona is being uplifted at a rate of 1–5 mm/yr
    D. Arizona is being uplifted at a rate of 5–10 mm/yr
    Answer = C for Arizona (page 556)

22

Energy Resources
from the Earth

1. Which of the following is a nonrenewable energy resource?
   A. solar
   B. methane
   C. hydroelectric
   D. coal
   Answer = D (page 562)

2. The entire amount of oil that may become available for use is called oil _____.
   A. reserves
   B. reservoirs
   C. resources
   D. traps
   Answer = C (page 562)

3. A coal deposit that is not economical to mine today would be considered part of our _____.
   A. coal reserves
   B. coal reservoirs
   C. coal resources
   D. none of these
   Answer = C (page 562)

4.   What is the leading source of energy used in the United States today?
          A.   oil
          B.   coal
          C.   natural gas
          D.   nuclear power
       Answer = A (page 563)

5.   The first oil well in the United States was drilled in _____.
          A.   1829
          B.   1859
          C.   1929
          D.   1959
       Answer = B (page 563)

Histogram showing percentages of various types of energy used in the United States today

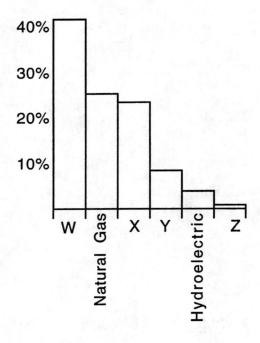

6.   Area W represents _____.
          A.   coal
          B.   solar
          C.   nuclear
          D.   oil
       Answer = D (page 563)

7.   Area X represents _____.
          A.   coal
          B.   solar
          C.   nuclear
          D.   oil
      Answer = A (page 563)

8.   Area Y represents _____.
          A.   coal
          B.   solar
          C.   nuclear
          D.   oil
      Answer = C (page 563)

9.   Area Z represents _____.
          A.   coal
          B.   solar
          C.   nuclear
          D.   oil
      Answer = B (page 563)

10.   Oil, coal, and natural gas supply approximately _____ percent of the energy used in the United States.
          A.   100
          B.   90
          C.   50
          D.   30
      Answer = B (page 564)

11.   Of all of the energy produced in the United states, what percentage is lost in distribution and inefficient use?
          A.   10
          B.   25
          C.   40
          D.   50
      Answer = C (page 564)

12.   Chemical reactions triggered by _____ transform organic material into hydrocarbons.
          A.   solar energy
          B.   hydroelectric energy
          C.   elevated temperatures
          D.   decomposition
      Answer = C (page 564)

13.   Energy resources derived from natural organic materials are called _____.
    A.   geothermal energy sources
    B.   fossil fuels
    C.   biomass
    D.   all of these
Answer = B (page 564)

14.   A permeable rock that contains hydrocarbon fluids and gases is called a(n) _____.
    A.   oil trap
    B.   source bed
    C.   oil reservoir
    D.   none of these
Answer = C (page 565)

15.   All oil traps contain _____.
    A.   an impermeable layer
    B.   an anticline
    C.   a fault
    D.   all of these
Answer = A (page 565)

16.   Which of the following is least likely to contain an oil trap?
    A.   an anticline
    B.   a fault
    C.   natural stratigraphy
    D.   a syncline
Answer = D (page 565)

Cross section of an oil trap. X, Y, and Z represent three distinct fluid layers within the permeable reservoir rock.

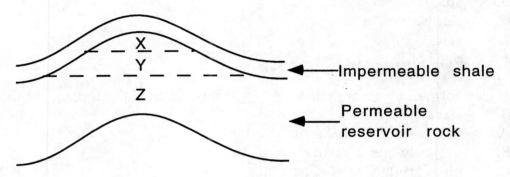

17.   The oil trap shown here is a(n) _____.
       A.   anticlinal trap
       B.   fault trap
       C.   stratigraphic trap
       D.   salt dome trap
       Answer = A (page 565)

18.   Layer X is most likely to be reservoir rock containing _____ in the pore space.
       A.   syncrude
       B.   water
       C.   natural gas
       D.   oil
       Answer = C (page 565)

19.   Layer Y is most likely to be reservoir rock containing _____ in the pore space.
       A.   syncrude
       B.   water
       C.   natural gas
       D.   oil
       Answer = D (page 565)

20.   Layer Z is most likely to be reservoir rock containing _____ in the pore space.
       A.   syncrude
       B.   water
       C.   natural gas
       D.   oil
       Answer = B (page 565)

21.   Which of the following rock types would make the best oil reservoir?
           A.   granite
           B.   shale
           C.   sandstone
           D.   salt
       Answer = C (page 565)

22.   In an oil trap formed by an anticline, _____ accumulates at the top, ___ in the mid-
       dle, and _____ at the bottom.
           A.   natural gas ... oil ... groundwater
           B.   groundwater ... oil ... natural gas
           C.   oil ... groundwater ... natural gas
           D.   oil ... natural gas ...groundwater
       Answer = A (page 565)

23.   Two-thirds of the world's known oil reserves are located in _____.
           A.   Siberia
           B.   the Gulf of Mexico and Caribbean
           C.   the Middle East
           D.   Indonesia
       Answer = C (page 566)

24.   The world has the least amount of which of the following types of fuel?
           A.   oil
           B.   coal
           C.   uranium
           D.   there are roughly equal amounts of each of these fuels
       Answer = A (page 568)

25.   At the current rate of world use, the remaining oil will be depleted in approximately
       _____ years.
           A.   25
           B.   100
           C.   400
           D.   2000
       Answer = B (page 568)

26.   Which of the following statements is true?
           A.   In the United States oil production is greater than oil consumption.
           B.   In the United States oil production roughly equals oil consumption.
           C.   In the United States oil production is less than oil consumption.
           D.   The United States does not produce any oil.
       Answer = C (page 568)

27.   A quad is _____.
            A.   a measure of the energy that can be extracted from a given amount of fuel
            B.   equal to 400,000 calories
            C.   equal to 1 BTU (British Thermal Unit)
            D.   all of the above
        Answer = A (page 568)

28.   Most of the natural gas used in the United States is consumed by _____.
            A.   industry
            B.   residential use
            C.   electric utilities
            D.   transportation
        Answer = A (page 569)

29.   Burning of which of the following fuels produces the least amount of carbon dioxide per unit of energy?
            A.   coal
            B.   oil
            C.   natural gas
            D.   all of these produce the same amount of carbon dioxide
        Answer = C (page 569)

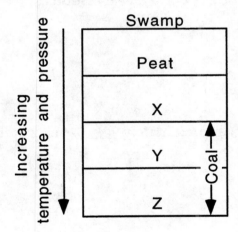

30.   Layer X is _____.
            A.   anthracite
            B.   bituminous coal
            C.   lignite
            D.   tar
        Answer = C (page 570)

31.   Layer Y is _____.
          A.   anthracite
          B.   bituminous coal
          C.   lignite
          D.   tar
      Answer = B (page 570)

32.   Layer Z is _____.
          A.   anthracite
          B.   bituminous coal
          C.   lignite
          D.   tar
      Answer = A (page 570)

33.   In addition to increasing pressure and temperature, the arrow to the left of the diagram represents increasing _____.
          A.   metamorphism
          B.   heat value
          C.   carbon content
          D.   all of the above
      Answer = D (page 570)

34.   Which country contains approximately 50% of the world's coal resources?
          A.   United States
          B.   China
          C.   Canada
          D.   the former Soviet Union
      Answer = D (page 571)

35.   In the United States, coal resources should last approximately _____ years at the current rate of use.
          A.   25
          B.   100
          C.   400
          D.   2,000
      Answer = C (page 571)

36.   Oil derived from coal, oil shales, or tar sand is called _____.
          A.   natural gas
          B.   biomass
          C.   syncrude
          D.   none of these
      Answer = C (page 571)

37.   Which of the following problems is associated with the burning of coal?
        A.   acid rain
        B.   carbon dioxide
        C.   ash with toxic metal impurities
        D.   all of these
        Answer = D (pages 571, 573)

38.   Nuclear energy is derived by _____.
        A.   the combustion of ^{235}U atoms
        B.   the fission of ^{235}U atoms
        C.   the fusion of ^{235}U atoms
        D.   the breaking of ^{235}U bonds
        Answer = B (page 573)

39.   Which of the following energy sources does not produce carbon dioxide?
        A.   oil
        B.   uranium
        C.   coal
        D.   natural gas
        Answer = B (page 573)

40.   How many nuclear power plants are there in the United States?
        A.   about 20
        B.   about 100
        C.   about 500
        D.   about 2,500
        Answer = B (page 574)

41.   Which of the following statements regarding the nuclear accident at Chernobyl in the Ukraine in 1986 is false?
        A.   Radioactive debris was blown into Scandinavia and western Europe.
        B.   Hundreds of square miles of land surrounding Chernobyl was contaminated and made uninhabitable.
        C.   Food supplies in many countries had to be purified in order to be consumed.
        D.   Excess deaths from cancer may be in the thousands over the next 40 years because of people exposed to the radiation.
        Answer = C (page 575)

42.   Most of the uranium ore in the United States is located in the _____.
          A.   Appalachian Mountains
          B.   Basin and Range Province
          C.   Colorado Plateau
          D.   Great Lakes region
       Answer = C (page 575)

43.   The primary barrier to using solar energy in the United States is that _____.
          A.   solar power is not technically feasible
          B.   solar power causes major pollution problems
          C.   solar power is not economically competitive with other energy sources
          D.   all of these
       Answer = C (page 576)

44.   Hydroelectric energy provides about ___ of the energy consumed annually in the
       United States.
          A.   4%
          B.   12%
          C.   30%
          D.   65%
       Answer = A (page 576)

45.   Solar energy stored in material such as wood, grain, sugar, and municipal waste is
       called _____.
          A.   fossil fuel
          B.   biomass
          C.   geothermal energy
          D.   natural gas
       Answer = B (page 576)

46.   What type of energy is derived from heated groundwater?
          A.   solar energy
          B.   geothermal energy
          C.   hydroelectric energy
          D.   nuclear energy
       Answer = B (page 577)

47.   The largest geothermal power plant in the United States is located near which city?
          A.   Chicago
          B.   Los Angeles
          C.   New York
          D.   San Francisco
       Answer = D (page 578)

48.   The Geysers produce enough electricity to meet what portion of the needs of San Francisco?
        A.   10%
        B.   25%
        C.   50%
        D.   100%
Answer = C (page 578)

49.   The world faces an energy crisis because _____.
        A.   the world demand for energy will increase
        B.   world oil production will peak and begin to decline
        C.   shortages and the resulting escalation of prices can shock the economic and political order
        D.   all of the above
Answer = D (page 579)

50.   Which of the following will not help to conserve energy?
        A.   better home insulation
        B.   changing from flourescent to incandescent lighting
        C.   more efficient refrigerators
        D.   more efficient automobile engines
Answer = B (page 579)

23

Mineral Resources
from the Earth

1.   Most metal ore deposits consist of _____.
            A.   native metals
            B.   metal oxides
            C.   metal sulfides
            D.   metal silicates
      Answer = C (page 586)

2.   In the United States the average per capita consumption of iron and steel is approxi-
      mately _____ per year.
            A.   1 pound
            B.   10 pounds
            C.   100 pounds
            D.   1000 pounds
      Answer = D (page 586)

3.   Quartz sand is a key ingredient in the manufacture of _____.
            A.   fertilizer
            B.   fiber optic cables
            C.   cement
            D.   steel
      Answer = B (page 586)

4.   The average abundance of iron in the Earth's crust is 5.8 weight %. In order to be economically mined, an iron ore deposit requires a concentration factor of _____.
      A.   0.5
      B.   2
      C.   10
      D.   100
Answer = C (page 587)

5.   Which of the following metals has the lowest economical concentration factor?
      A.   aluminum
      B.   copper
      C.   gold
      D.   iron
Answer = A (page 588)

6.   The United States imports approximately _____ of its aluminum.
      A.   25%
      B.   50%
      C.   75%
      D.   100%
Answer = D (page 589)

7.   Nearly 75% of which metal is currently recycled in the United States?
      A.   aluminum
      B.   copper
      C.   iron
      D.   lead
Answer = D (page 589)

8.   When was the U.S. Mining Act enacted?
      A.   1872
      B.   1912
      C.   1942
      D.   1982
Answer = A (page 590)

9.   In 1992, $6.2 billion was paid to the federal government as rents and royalties for the use of federal lands. What percentage of this federal income was paid by mineral-mining interests?
      A.   about 1%
      B.   about 10%
      C.   about 50%
      D.   about 90%
Answer = A (page 590)

10.   Which of the following statements regarding hydrothermal ore deposits is false?
     A.   Placers represent one type of hydrothermal ore deposit.
     B.   Hydrothermal ore deposits are commonly associated with shallow granitic intrusions.
     C.   Minerals precipitating from "black smokers" at mid-ocean spreading centers are an example of a modern hydrothermal ore deposit.
     D.   Most hydrothermal ore deposits consist of metallic sulfides.
Answer = A (page 592)

11.   What type of ore deposits form where minerals precipitate along the walls of faults and joints?
     A.   disseminated deposits
     B.   hydrothermal vein deposits
     C.   placer deposits
     D.   sedimentary deposits
Answer = B (page 592)

12.   Which of the following metals is not commonly found in ore deposits that form from hydrothermal solutions?
     A.   aluminum
     B.   lead
     C.   copper
     D.   zinc
Answer = A (page 592)

13.   Hydrothermal ore-bearing fluids form _____.
     A.   directly from a crystallizing magma
     B.   when groundwater comes into contact with a hot intrusion
     C.   both A and B
     D.   neither A nor B
Answer = C (page 592)

14.   In hydrothermal vein deposits, most metals are precipitated as _____.
     A.   native elements
     B.   silicates
     C.   oxides
     D.   sulfides
Answer = D (page 593)

15.   Which of the following ore deposits are disseminated deposits?
         A.   the iron deposits of the Great Lakes region
         B.   the copper deposits of the southwest U.S.
         C.   the gold placer deposits of California
         D.   the coal deposits of the Appalachian Mountains
         Answer = B (page 594)

16.   What is the major metal that is extracted from the large open-pit mines of the south-
         western U.S. and Chile?
         A.   iron
         B.   nickel
         C.   copper
         D.   zinc
         Answer = C (page 594)

17.   The most common copper ore mineral in the porphyry-copper deposits of the south-
         west U.S. is _____.
         A.   native copper
         B.   copper silicate
         C.   copper oxide
         D.   copper sulfide
         Answer = D (page 594)

18.   What ore mineral is the major source of lead?
         A.   sphalerite
         B.   pyrite
         C.   chalcopyrite
         D.   galena
         Answer = D (page 595)

19.   Which of the following mineral resources is not generally associated with
         mafic/ultramafic igneous rocks?
         A.   lead
         B.   platinum
         C.   chromium
         D.   diamonds
         Answer = A (page 595)

20. Which of the following metals is concentrated by crystals settling on the floor of a magma chamber?
    A. platinum
    B. lead
    C. gold
    D. iron
Answer = A (page 595)

21. Which of the following types of ore deposits contains rare elements, such as boron, lithium, and fluorine?
    A. ophiolites
    B. pegmatites
    C. placers
    D. porphyry deposits
Answer = B (page 595)

22. What type of rock is required to make cement and is an important building stone?
    A. limestone
    B. granite
    C. basalt
    D. quartz sandstone
Answer = A (page 596)

23. Diamonds are found in _____ igneous rocks, called kimberlites, that erupt from great depth.
    A. felsic
    B. intermediate
    C. mafic
    D. ultramafic
Answer = D (page 596)

24. Which of the following materials is not derived from sedimentary ore deposits?
    A. cement
    B. sand and gravel
    C. ceramics
    D. all of these are derived from sedimentary ore deposits
Answer = D (pages 596-597)

25. Evaporite deposits are not mined for _____.
    A. plaster
    B. iron
    C. fertilizer
    D. table salt
Answer = B (page 597)

26.   Iron ore deposits, such as occur in the Lake Superior region, are primarily _____ ore deposits.
       A.   igneous
       B.   hydrothermal
       C.   sedimentary
       D.   disseminated
Answer = C (page 597)

27.   The world's major iron ore deposits were formed during _____ time.
       A.   Precambrian
       B.   Paleozoic
       C.   Mesozoic
       D.   Cenozoic
Answer = A (page 597)

28.   The rich copper-bearing Kupferschiefer deposits of Germany formed in _____ rocks.
       A.   volcanic
       B.   plutonic
       C.   sedimentary
       D.   high-grade metamorphic
Answer = C (page 597)

29.   Copper ore deposits are _____.
       A.   hydrothermal ore deposits
       B.   igneous ore deposits
       C.   sedimentary ore deposits
       D.   all of these
Answer = D (page 597)

30.   The Mother Lode ore deposit in California, discovered in the middle of the 19th century, is rich in _____.
       A.   iron
       B.   copper
       C.   gold
       D.   diamonds
Answer = C (page 598)

31.   What type of ore deposits have been concentrated by the mechanical sorting action of river currents?
       A.   igneous ore deposits
       B.   placer deposits
       C.   hydrothermal ore deposits
       D.   disseminated ore deposits
Answer = B (page 598)

Schematic cross section

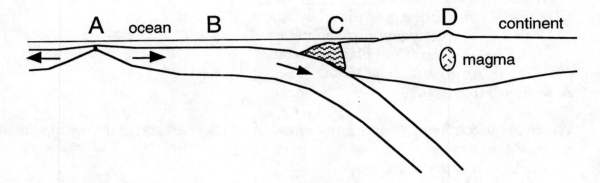

32. In what plate tectonic setting are copper porphyry deposits located?
    A. plate tectonic setting A
    B. plate tectonic setting B
    C. plate tectonic setting C
    D. plate tectonic setting D
    Answer = D (page 600)

33. Rich copper-, lead-, and zinc-sulfide deposits, like those of Cyprus, form in which plate tectonic setting?
    A. plate tectonic setting A
    B. plate tectonic setting B
    C. plate tectonic setting C
    D. plate tectonic setting D
    Answer = A (page 600)

34. Manganese nodules forms in which region?
    A. plate tectonic setting A
    B. plate tectonic setting B
    C. plate tectonic setting C
    D. plate tectonic setting D
    Answer = B (page 600)

35. Ophiolites and their associated ore deposits are located in which plate tectonic setting?
    A. plate tectonic setting A
    B. plate tectonic setting B
    C. plate tectonic setting C
    D. plate tectonic setting D
    Answer = C (page 600)

36.   The copper ore deposits of Cyprus, which played an important role in the economy of ancient Greece, formed _____.
        A.   at a convergent plate boundary
        B.   at a divergent plate boundary
        C.   at a transform plate boundary
        D.   in an intraplate setting
Answer = B (page 600)

37.   Which of the following metals is not found in "black smokers" that form at spreading centers?
        A.   copper
        B.   aluminum
        C.   zinc
        D.   iron
Answer = B (page 600)

38.   Rich, hydrothermal ore deposits composed of copper, lead, and zinc sulfides are associated with _____.
        A.   placers
        B.   kimberlites
        C.   evaporites
        D.   ophiolites
Answer = D (page 600)

39.   Nodules rich in _____ are found on the deep seafloor away from plate boundaries.
        A.   gold
        B.   aluminum
        C.   manganese
        D.   potassium
Answer = C (page 601)

North America

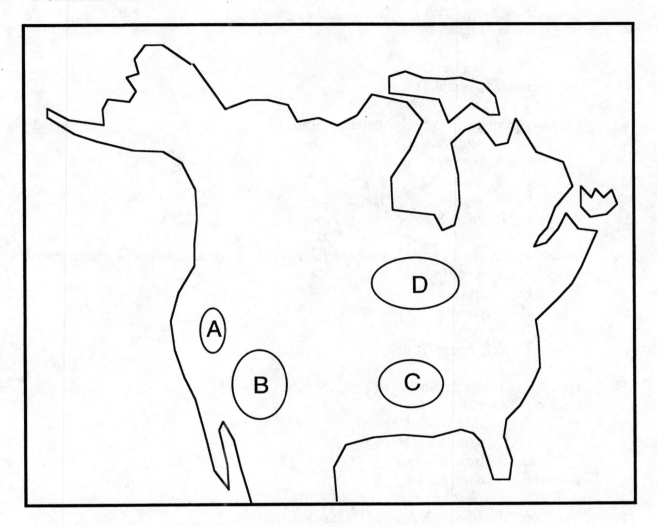

40.   Which region provides most of the iron ore for North America?

         A.   region A
         B.   region B
         C.   region C
         D.   region D

Answer = D (page 602)

41.   In which region is the Mother Lode located?

         A.   region A
         B.   region B
         C.   region C
         D.   region D

Answer = A (page 602)

42.   Which region contains lead and zinc deposits hosted by sedimentary rocks?
        A.   region A
        B.   region B
        C.   region C
        D.   region D
    Answer = C (page 602)

43.   Large copper porphyry deposits are found in _____.
        A.   region A
        B.   region B
        C.   region C
        D.   region D
    Answer = B (page 602)

44.   What type of metal is concentrated in regions of intense chemical weathering?
        A.   aluminum
        B.   iron
        C.   copper
        D.   gold
    Answer = A (page 603)

45.   Iron ore consists primarily of _____.
        A.   native iron
        B.   iron oxides
        C.   iron sulfides
        D.   iron silicates
    Answer = B (page 603)

46.   The Republic of South Africa has large deposits of _____.
        A.   diamonds
        B.   platinum
        C.   gold
        D.   all of these
    Answer = D (page 603)

47.   What is the chemical formula for covellite, an important copper ore mineral?
        A. CuO
        B. CuS
        C. $CuSiO_3$
        D. $CuCO_3$
    Answer = B (page 603)

48.   Which of the following is not a correct match of a metal and the process by which the metal ore deposit forms?
          A.   gold --- concentration in placers by river currents
          B.   copper --- precipitation from hydrothermal fluids associated with granitic intrusions
          C.   lead --- precipitation from hydrothermal fluids in sedimentary rocks
          D.   aluminum   --- precipitation in shallow Precambrian seas
        Answer = D (page 603)

49.   Over the past forty years, the percentage of nonfuel minerals produced and consumed in the United States has _____ relative to the rest of the world.
          A.   increased rapidly
          B.   increased slowly
          C.   remained essentially constant
          D.   decreased
        Answer = D (page 605)

50.   The "economic exclusion zone" refers the exclusive rights to mineral deposits in the offshore area within _____ miles of a country's coast.
          A.   3
          B.   12
          C.   75
          D.   200
        Answer = D (page 607)

24

Earth Systems and Cycles

1. The greenhouse effect is caused by _____.
    A. CO_2 and H_2O vapor which trap heat radiating from the Earth's surface
    B.   oceans which trap heat radiating from the Earth's seafloors
    C.   heating of homes and businesses, which releases an excess of heat into the atmosphere
    D.   too many plants on the surface of the Earth, which do not allow sufficient cooling of the ground surface
    Answer = A (page 613)

2. The process by which green organisms use chlorophyll and the energy from sunlight to make carbohydrates is called _____.
    A.   respiration
    B.   greening
    C.   photosynthesis
    D.   carbohydration
    Answer = C (page 614)

$$\text{carbohydrate + oxygen} \xrightarrow{\text{release of chemical energy}} \text{carbon dioxide + water}$$

3.   This chemical reaction represents _____.
        A.   respiration
        B.   photosynthesis
        C.   global warming
        D.   the greenhouse effect
    Answer = A (page 614)

4.   In the early Earth's oxygen depleted atmosphere, weathering and other surface processes would have _____.
        A.   operated more rapidly
        B.   operated less rapidly
        C.   operated much the same as today
        D.   operated more rapidly initially, then stopped abruptly
    Answer = C (page 614)

5.   For each atom of carbon taken from carbon dioxide during photosynthesis, _____ molecules of oxygen (O_2) are formed.
        A.   one
        B.   two
        C.   three
        D.   four
    Answer = A (page 614)

6.   If photosynthesis and respiration were in perfect balance globally, _____.
        A.   all organic matter would be used up in the respiration process
        B.   the carbon dioxide content of the atmosphere would be double the current value
        C.   the atmosphere would contain twice as much oxygen as carbon dioxide
        D.   oxygen levels would slowly decrease
    Answer = A (page 615)

7.   The excess of photosynthesis over respiration is due to _____
        A.   the overabundance of plant matter on the Earth
        B.   new plant diseases, which stunt plant growth
        C.   the increasing CO_2 content of the atmosphere
        D.   the burial of plant matter
    Answer = D (page 615)

8. _____ molecule(s) of oxygen will be released into the atmosphere for each carbon atom that is buried in organic matter, because respiration cannot take place.
    A. One
    B. Two
    C. Ten
    D. Twenty
    Answer = A (page 615)

9. Oxygen levels have built up in Earth's atmosphere over many years due to _____.
    A. degassing of meteorites
    B. plant burial
    C. volcanic degassing
    D. formation of ocean water
    Answer = B (page 615)

10. The stratosphere is the _____.
    A. lower atmosphere
    B. mid-atmosphere
    C. upper atmosphere
    D. vacuum above the upper atmosphere
    Answer = C (page 615)

11. Oxygen molecules which diffuse into the upper atmosphere are transformed into _____ by solar radiation.
    A. carbon dioxide
    B. water vapor
    C. ozone
    D. water
    Answer = C (page 615)

12. The stratospheric ozone layer provides the Earth with some protection from _____.
    A. solar wind
    B. ultraviolet rays
    C. solar x-rays
    D. oxidation
    Answer = B (page 615)

13.   Land plants and animals evolved _____.
        A.   approximately 4 billion years ago
        B.   between 3.5 and 1.5 billion years ago
        C.   between 1.5 and 0.45 years ago
        D.   between 0.45 billion years ago and the present
        Answer = D (page 616)

14.   The explosion of life forms in the late Precambrian and Cambrian was probably stimulated by _____.
        A.   a decrease in harmful bacteria strains
        B.   an increase in bacterial food sources
        C.   a rise in oxygen levels
        D.   a rise in nitrogen levels
        Answer = C (page 616)

15.   When the inflow of an element into a reservoir equals the outflow, the average amount of time an atom of that element spends in the reservoir is the _____.
        A.   travel time
        B.   equilibrium time
        C.   residence time
        D.   rotation time
        Answer = C (page 618)

16.   The residence time of sodium in the ocean is approximately _____ years.
        A.   50 million
        B.   850,000
        C.   5,000
        D.   85
        Answer = A (page 618)

17.   The residence time of iron in the ocean is approximately _____ years.
        A.   1 million
        B.   100,000
        C.   1,000
        D.   100
        Answer = D (page 618)

Ocean-atmosphere exchange

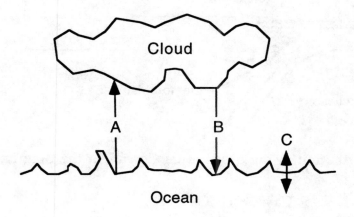

18.   Arrow A refers to _____.
               A.   evaporation
               B.   runoff
               C.   particle settling
               D.   precipitation
        Answer = A (page 619)

19.   Arrow B refers to _____.
               A.   evaporation
               B.   runoff
               C.   gas exchange
               D.   precipitation
        Answer = D (page 619)

20.   Arrow C refers to _____.
               A.   evaporation
               B.   particle settling
               C.   gas exchange
               D.   precipitation
        Answer = C (page 619)

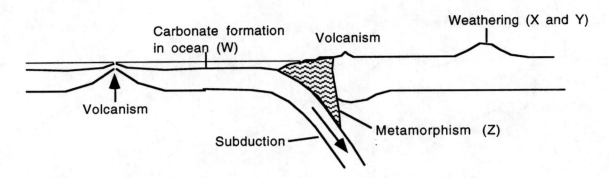

21.   The equation for carbonate formation in the oceans (reaction W in the diagram) is
_____.

        A. $CO_2 + H_2O + CaCO_3 \rightarrow Ca^{2+} + 2HCO_3^-$

        B. $2CO_2 + H_2O + CaSiO_3 \rightarrow Ca^{2+} + 2HCO_3^- + SiO_2$

        C. $2HCO_3^- + Ca^{2+} \rightarrow CaCO_3 + CO_2 + H_2O$

        D. $CaCO_3 + SiO_2 \rightarrow CaSiO_3 + CO_2$

Answer = C (page 619)

22.   The equation for silicate weathering (reaction X in the diagram) is _____.

        A. $CO_2 + H_2O + CaCO_3 \rightarrow Ca^{2+} + 2HCO_3^-$

        B. $2CO_2 + H_2O + CaSiO_3 \rightarrow Ca^{2+} + 2HCO_3^- + SiO_2$

        C. $2HCO_3^- + Ca^{2+} \rightarrow CaCO_3 + CO_2 + H_2O$

        D. $CaCO_3 + SiO_2 \rightarrow CaSiO_3 + CO_2$

Answer = B (page 619)

23.   The equation for carbonate weathering (reaction Y in the diagram) is
_____.

        A. $CO_2 + H_2O + CaCO_3 \rightarrow Ca^{2+} + 2HCO_3^-$

        B. $2CO_2 + H_2O + CaSiO_3 \rightarrow Ca^{2+} + 2HCO_3^- + SiO_2$

        C. $2HCO_3^- + Ca^{2+} \rightarrow CaCO_3 + CO_2 + H_2O$

        D. $CaCO_3 + SiO_2 \rightarrow CaSiO_3 + CO_2$

Answer = A (page 619)

24. The equation for metamorphic breakdown of carbonate (reaction Z in the diagram) is _____.

  A. $CO_2 + H_2O + CaCO_3 \rightarrow Ca^{2+} + 2HCO_3^-$

  B. $2CO_2 + H_2O + CaSiO_3 \rightarrow Ca^{2+} + 2HCO_3^- + SiO_2$

  C. $2HCO_3^- + Ca^{2+} \rightarrow CaCO_3 + CO_2 + H_2O$

  D. $CaCO_3 + SiO_2 \rightarrow CaSiO_3 + CO_2$

Answer = D (page 619)

25. Which of the following reactions and processes release CO^2 into the atmosphere?
  A. reactions W and X
  B. reactions W and X and volcanism
  C. reactions Y and Z
  D. reactions Y and Z and volcanism

Answer = D (page 619)

26. Which of the following processes does not contribute to carbon dioxide entering the atmosphere?
  A. volcanism
  B. sedimentation of calcium carbonate
  C. weathering of carbonate rocks
  D. metamorphism

Answer = C (page 620)

27. The biosphere contains _____.
  A. the upper, middle, and lower atmosphere
  B. the middle and lower atmosphere and the surface of the crust
  C. the lower atmosphere, the surface of the crust and the hydrosphere
  D. the surface of the Earth, the hydrosphere, and the asthenosphere

Answer = C (page 620)

28. The amount of radiation from the Sun reaching the upper atmosphere is called the

_____.
  A. solar density
  B. solar constant
  C. solar emission
  D. radiation limit

Answer = B (page 621)

29.   Sulfuric acid aerosols formed as a result of volcanic eruptions affect the atmosphere by _____.

      A.   absorbing oxygen released from photosynthesis
      B.   preventing efficient air circulation in the lower atmosphere
      C.   preventing heat in the atmosphere to escape
      D.   absorbing solar radiation

Answer = D (page 621)

30.   The mass extinction that occurred between the Permian and Triassic Periods wiped out _____ of the marine species and _____ of land vertebrates.

      A.   30% ... 10%
      B.   50% ... 30%
      C.   70% ... 50%
      D.   90% ... 70%

Answer = D (page 622)

31.   Which of the following are not observed in the geologic record at the time of the mass extinction between the Permian and Triassic Periods?

      A.   a sea-level drop
      B.   an ice-sheet expansion
      C.   an increase in sodium content in the oceans
      D.   an interval of acid rain

Answer = C (page 622)

32.   The sulfur dioxide gas from the eruption of a superplume could affect life on Earth by _____.

      A.   forming a sulfuric acid haze
      B.   contributing to acid rain
      C.   forming sulfur dioxide aerosols which block sunlight
      D.   all of the above

Answer = D (page 623)

33.   The leading theory of the cause of the Cretaceous-Tertiary mass extinction is _____.

      A.   a superplume eruption
      B.   a bolide impact
      C.   a deadly virus strain
      D.   none of the above

Answer = D (page 623)

34.   Evidence of a bolide impact would not include which of the following?
        A.   flood basalt deposits
        B.   shock-metamorphosed quartz
        C.   high levels of iridium in sediments
        D.   an impact crater
        Answer = A (page 624)

35.   The idea that human activities could result in emissions that could alter the chemistry
        of the atmosphere is called _____.
        A.   global warming
        B.   global change
        C.   global emission
        D.   global ramification
        Answer = B (page 626)

36.   Which of the following is not a likely result of global change?
        A.   climate change due to an enhanced greenhouse effect
        B.   increased exposure to ultraviolet rays due to stratospheric ozone deple-
                tion
        C.   increased variety of life forms due to competitors extinctions
        D.   an overburdening of many Earth systems due to overpopulation
        Answer = C (page 626)

37.   Which of the following substances is not a greenhouse gas?
        A.   methane
        B.   nitrous oxide
        C.   chlorofluorocarbon
        D.   sulfur dioxide
        Answer = D (page 626)

38.   By the end of the next century temperatures are predicted to rise approximately
        _____ due to global warming.
        A.   less than 1°C
        B.   1 to 3.5°C
        C.   3.5 to 10°C
        D.   greater than 10°C
        Answer = B (page 626)

39.   If the continental glaciers as well as the Greenland and Antarctic ice sheets melt, sea level could rise as much as _____.
    A.   6.5 centimeters
    B.   65 centimeters
    C.   6.5 meters
    D.   65 meters
Answer = D (page 626)

40.   If all human activities that generated carbon dioxide were to stop, how long would it take for atmospheric carbon dioxide to return to its preindustrial level?
    A.   2 months
    B.   2 years
    C.   20 years
    D.   200 years
Answer = D (page 626)

41.   Which of the following will not reduce the emissions of greenhouse gases?
    A.   using coal rather than oil and gasoline
    B.   using nuclear power redesigned for safety
    C.   using fuel made from renewable resources such as wood and grain
    D.   using vehicles powered by electricity
Answer = A (page 626)

42.   What is the chemical formula for ozone?
    A. O
    B. O_2
    C. O_3
    D. O_4
Answer = C (page 627)

43.   Which of the following is not a use of CFCs (chlorofluorocarbons)?
    A.   fluorescent lighting
    B.   refrigerants
    C.   propellants
    D.   cleaning solvents
Answer = A (page 628)

44.   When sunlight reacts with CFCs (chlorofluorocarbons), _____ is released.
    A.   chlorine
    B.   fluorine
    C.   carbon
    D.   carbon dioxide
Answer = A (page 628)

45.   A large hole in the ozone layer was found over _____.
         A.   the North Pole
         B.   Siberia
         C.   Brazil
         D.   Antarctica
Answer = D (page 628)

46.   Acid rain is composed of _____ acids.
         A.   sulfuric and oxalic
         B.   sulfuric and nitric
         C.   oxalic and hydrochloric
         D.   hydrochloric and nitric
Answer = B (page 629)

47.   Sulfur dioxide is emitted from power plants that burn coal containing large amounts
        of the mineral _____.
         A.   muscovite
         B.   pyrite
         C.   biotite
         D.   hematite
Answer = B (page 629)

48.   In Canada alone, acid rain causes _____ worth of damage to buildings and
        monuments each year.
         A. $1 million
         B. $10 million
         C. $100 million
         D. $1 billion
Answer = D (page 629)

49.   Which area in the United States has the most acidic rain?
         A.   northwest U.S.
         B.   southwest U.S.
         C.   northeast U.S.
         D.   southeast U.S.
Answer = C (page 630)

50.   Economic growth and improved standards of living which can last indefinitely and
        are environmentally kind are called _____.
         A.   global development
         B.   global standards
         C.   sustainable standards
         D.   sustainable development
Answer = D (page 631)